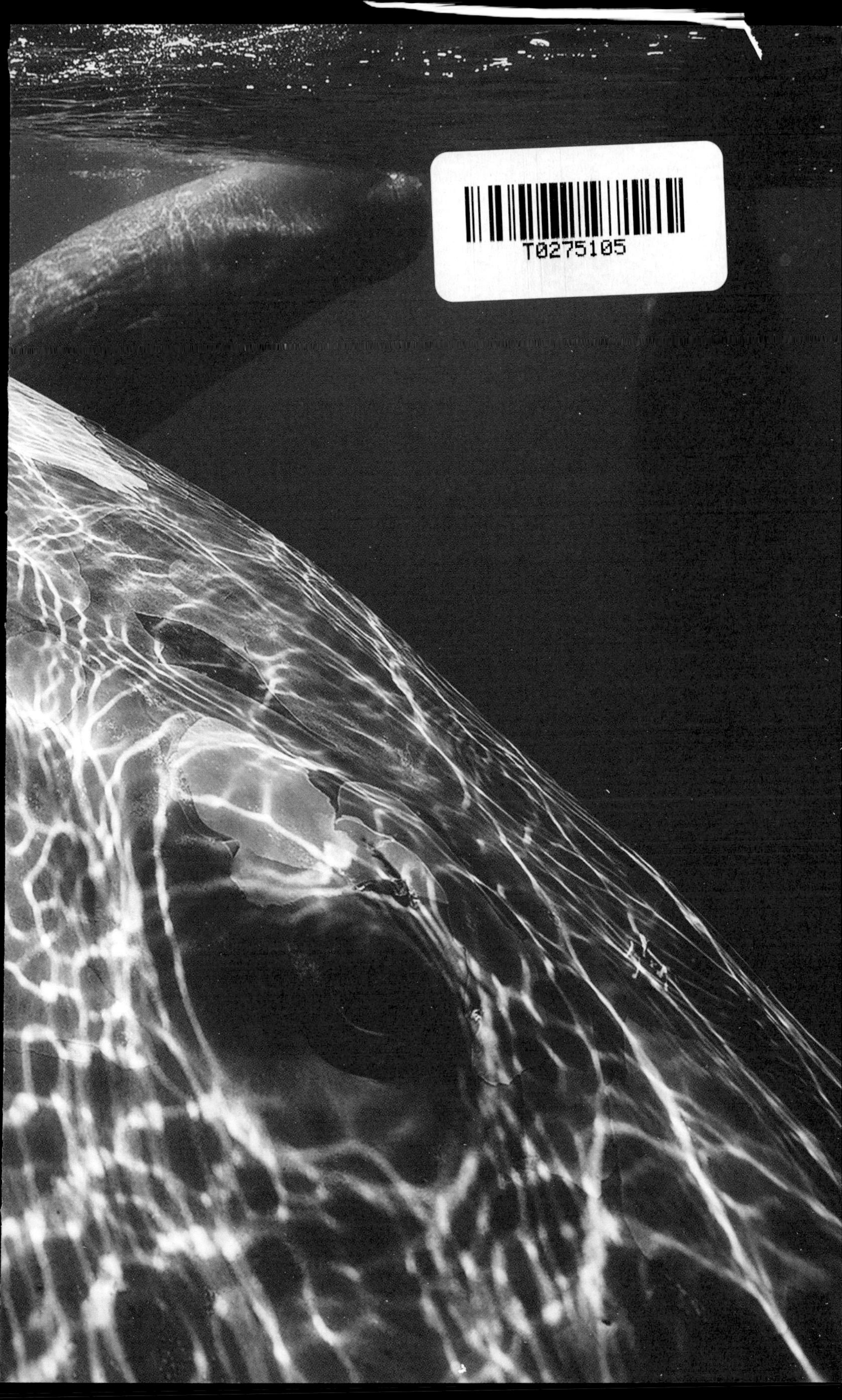
T0275105

WHALE MUSIC

Also by David Rothenberg

*Is It Painful to Think?*

*Hand's End*

*Sudden Music*

*Blue Cliff Record*

*Always the Mountains*

*Why Birds Sing*

*Thousand Mile Song*

*Bug Music*

*Nightingales in Berlin*

*The Possibility of Reddish Green*

*From This World, Another*

[with Stephen Nachmanovitch]

*Invisible Mountains*

# WHALE MUSIC

## Thousand Mile Songs in a Sea of Sound

DAVID ROTHENBERG

*featuring photos by*
*Wade and Robyn Hughes*

**Terra Nova Press**
NEWARK CALLICOON MATSALU

2023

ISBN: 978-1-949597-25-7

Library of Congress Control Number: 2022944001


a revised and updated version of
David Rothenberg, *Thousand Mile Song: Whale Music in a Sea of Sound*
(Basic Books, 2008)

**Terra Nova Press**
NEWARK CALLICOON MATSALU

Editor-in-Chief: Evan Eisenberg
Editor and proofreader: Tyran Grillo
Book design: Martin Pedanik
Set in Spectral

printed by Tallinn Book Printers, Tallinn, Estonia

1 2 3 4 5 6 7 8 9 10

**www.terranovapress.com**

Distributed by the MIT Press, Cambridge, Massachusetts and London, England

*To all the great cetaceans of the sea,*
*as we listen—*
*may we help you*
*to survive*

# CONTENTS

# FOREWORD

To understand the seeds of David Rothenberg's *Thousand Mile Song* (2008), one need only read his sixth book, *Why Birds Sing* (2005), a dazzling exploration that begins in the National Aviary in Pittsburgh where he responded to the singing of a white-crested laughingthrush by playing his clarinet, creating a duet. Five years later, the tale ends in New South Wales, where an ancient Albert's lyrebird named George is still singing his 27-part song even though no females are in the area:

> George arches his lyre-shaped glimmering tail feathers right over his head into a curved dome of gossamer maroon. A single red back feather sticks straight up from his behind, a sudden bright surprise.... First, a *sneep* of the crimson rosella parrot, then the *plink chee chee chee chee* of the tiny yellow robin, then a high *pink* of a green catbird, then the harsh *braaaa* of the paradise riflebird. Next, a phrase of descending whistles and scratches from the satin bowerbird, that famous species who so loves blue. *Ba bo de poo traaaaaawh tete pu traaaawhh aaaarrrhh.* In our language it looks like nonsense but in George's voice it is strictly controlled: *Chik. Arrh. .... Broo ah ha ha,* the laughing kookaburra. *Cree* of the Lewin's honeyeater. *Plink chee chee chee chee* of the yellow robin again. *Aaaarrrhh* of the bowerbird. Parrot *chiks.* Flapping wings emulated as song, then faint burbles of a white-browed scrub wren. *Mraaah* of the riflebird. A break and a turn for the territorial call: *Breaap, booua, bwe, ba boo pu tee!* Back again to the rosella *sneep*. The cycle continues, with only slight variation.

David is a philosopher and clarinetist who traveled the Earth from Tortola to Karelia to Samaná to hear diverse songs of whales and to revision them in colored forms to give them three-dimensionality. He has also closely followed the literature of more than 200 papers that contribute to our growing understanding, beginning with my cover paper with Roger Payne for *Science* (1971), describing the discovery and documentation of this six-octave, six-part song from recordings in the 1960s by Frank Watlington in Bermuda.

What Rothenberg has written, and carefully annotated, is an exciting read, a sprightly narrative that pulls the reader in and holds her in thrall—like Coleridge's *The Rime of the Ancient Mariner* or Ishmael's enrapturing tale. Frank Zelko of the University of Vermont has written that the euphoric response by folks worldwide to the arias of the humpback male is what prompted the International Whaling Commission to create a so-called ban on whaling in 1986.

But the slaughter continues. The Japanese, while allegedly killing only sperm whales and minke whales, are killing endangered and threatened whales whenever they encounter them, including blues, fins, southern rights, sei, and humpbacks, as we know from taking tiny samples of whale meat in the Tokyo market that may be analyzed nearby in one's hotel room.

For 2022, David has revised his book and retitled it *Whale Music: Thousand Mile Songs in a Sea of Sound*, updating his story with new developments in the science and art of whale song and how humans have reacted to it over the decade and a half since the publication of the first edition.

Here are a few notable whale developments of recent years:

The film documentary, *The Cove* (2009), directed by Louie Psihoyos, won an Academy Award showing the bloody dolphin slaughter at night in the coastal community of Taiji, which is not an anomaly but rather a persisting tragedy of horrific consequence. Estimates suggest that the death drives kill 23,000 dolphins every year.

The most astonishing publication of the past decade to open wider our perception of the elusive tribe of Cetacea is Hal Whitehead and Luke Rendell's book *The Cultural Lives of Whales and Dolphins* (2014). With meticulous care by two acoustical biologists who document the elegance and nuance of sonic phonations of sperm whales and other species, we begin to see further the scientific underpinning of the book, *Mind in the Waters: A Book to Celebrate the Consciousness of Whales and Dolphins* (1974), which suggests a level of awareness in our fellow marine mammals that should not be surprising when you know their behaviors, the elaborate structures of their brains, and sonic repertoires. The book focuses on the evolution of social learning, which opens an astounding porthole into the fascinating cultures beneath the waves. Those two noted this interesting fact from the history of human study of whale music:

> Payne and McVay's paper in the influential scientific journal *Science* includes the statement: "Humpback whales emit a series of surprisingly beautiful sounds." Rothenberg, a philosopher and musician, notes—rather sadly, it seems—that no scientists had used the word "beauty" in a technical paper about whale song before. Or since. We, as whale scientists, know why. But Payne and McVay had the guts to include the b word. Indeed, they did much more than note the song's beauty—they used it.

For a long time now, *The Guardian* in the UK has been the lead reporter—followed by *The New York Times*—on the continuing assault on whales through hunting, PCB concentrations in their bodies along with other commercial contaminants, the unending high decibel sounds of ships, seismic soundings, and oil drilling explosions, food deprivation (salmon among resident orca pods in the northeast Pacific, even krill for baleen whales in the Antarctic), and the irresolution of the International Whaling Commission in the face of a 70-person delegation from Japan peddling a "Way Forward" plan to reinstate commercial whaling. That occurred in Florianopolis, Brazil, in September 2018.

Nick Pyenson nails the dim outlook in *Spying on Whales* (2018): "Today the all-encompassing influence of climate change represents the greatest human impact—in magnitude and rate—that we have yet had on the lives of whales, or any other creature on Earth." The lead country arguing for protection of whales in the twenty-first century after decades of over-exploitation is Australia, followed by New Zealand and the European Union. The rising voices of articulate women are cited repeatedly in the press.

Sadly, and tragically, Norway was in cahoots with Japan by taking 432 minke whales in 2018, while bumping up the quota to 1,200+. Iceland took 141 fin whales in 2018. These whales are all shipped to Japan, where the country has largely lost its appetite for eating whale meat.

We had thought that as we learn more about this wondrous natural world, including the still mysterious whales of the deep—notably, the surprisingly beautiful arias of humpback whales—that that knowledge would slow or even stop our assault on the biological underpinnings of the human experiment. That was boldly argued in the late Edward O. Wilson's book *Half-Earth* (2016). Thus far, we are not becoming the guardians and stewards of this precious, irreplaceable garden of life.

This book, *Whale Music*, and its predecessors, *Why Birds Sing*, *Survival of the Beautiful*, and *Nightingales in Berlin*, are a continuing spur for us to gain the will to keep and treasure life here on Earth in all its diversity.

Scott McVay
Director Emeritus, Geraldine R. Dodge Foundation
Author, *Surprise Encounters with Artists and Scientists*, *Whales and Other Living Things*, and *Whale Songs and Other Exuberances*

October 2021, Princeton, NJ

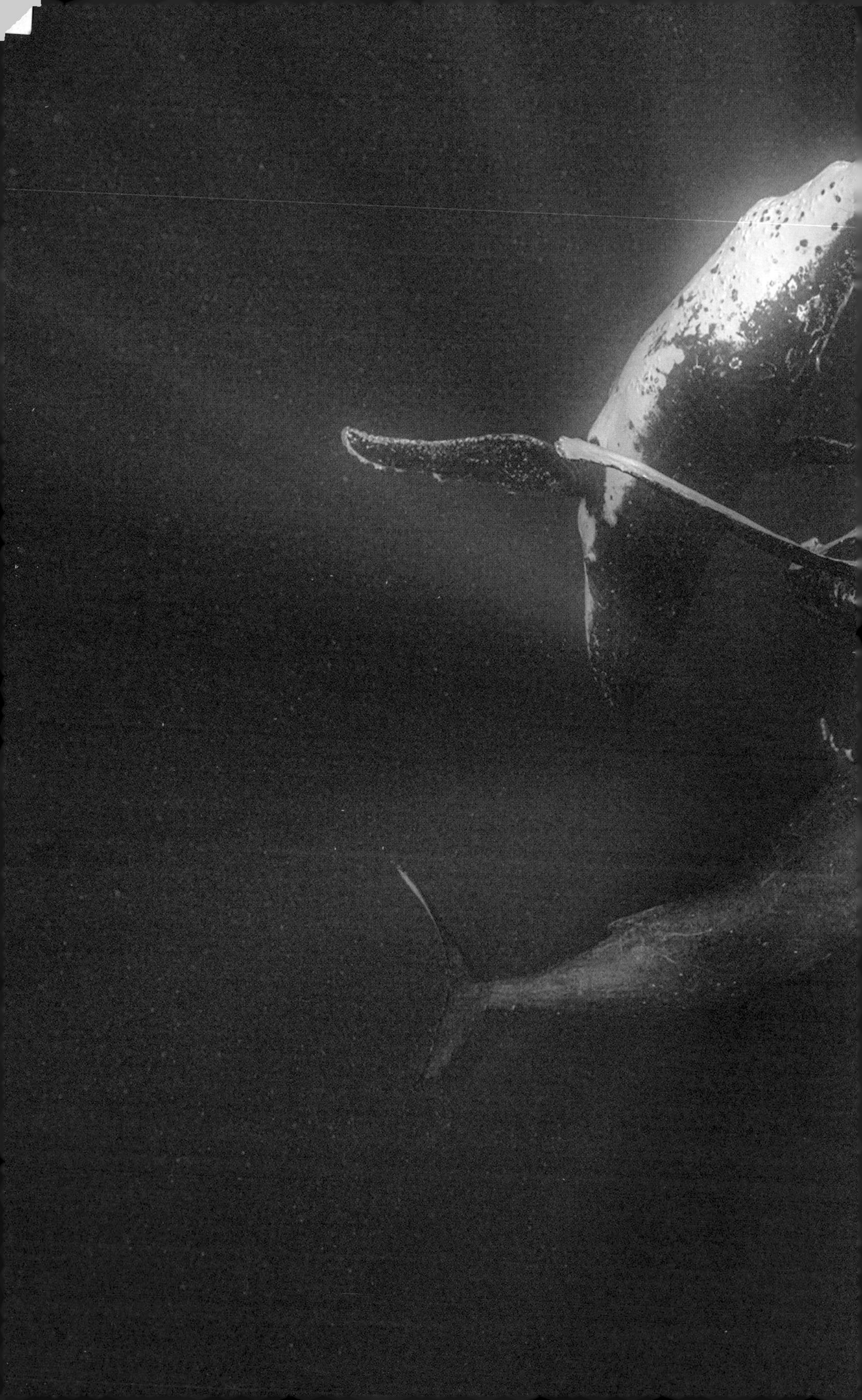

# 1

## WE DIDN'T KNOW, WE DIDN'T KNOW

### Whale Song Hits the Charts

Every morning Paul Knapp sets out from the shores of Tortola in hopes of recording a humpback whale song better than the one he heard on Valentine's Day, 1992.

"I remember that day well," nods Paul, looking up at the sky. "It was the fourteenth of February, and I was all alone. I went slowly and with respect, to the spot I always go to listen at the mouth of the bay. I didn't even see the whale. I think he was used to me by then and used to the sound of my engine. The whole moment made sense." Tall, tanned, in his mid-fifties, he looks like a man patient enough to spend years searching for the thing that matters most. "It's never been quite like that again. But I keep coming back—waiting, listening."

Knapp spends the winter months camped out on the beach at Brewer's Bay, an isolated cove in the Virgin Islands. The road to the place is still an improbable paved-over goat path. People here look pleased to have found a way to get away from it all. Some have come every winter for thirty years and say it hasn't changed a bit.

Paul stays here because of the whales, and the joy he finds in taking others out to hear their beautiful sounds. People sunning themselves on the beach get a surprise vacation treat when Knapp walks right up to them and says, "Hey, wanna come out on my boat and hear some whale songs?"

He pulls his tiny Caribe raft into the blue-green water from the perfect sandy arc of the beach and motors out to the mouth of the cove—as far as he can go before the whitecaps get too rough for a rubber boat. He unpacks his equipment from a watertight box: an underwater microphone (known as a hydrophone), with a cable a hundred feet long. Paul drops this over the side of the boat only after tying it off to a guyline lest the thing disappear and coil swiftly down to the bottom. He plugs it into a little battery-powered boom box and gets ready to listen.

Sound underwater works in an entirely different way than above the surface. It travels about fifteen hundred meters per second, five times the speed it moves in air. Sounds travel much farther submerged than above the surface: the lower the pitch, the farther they go. There is also the constant background noise of waves, the din of endless shipping vessels from motorboats to tankers, and seismic rumblings from far inside the earth. But the loudest thing we hear today is shrimp.

The sound is surprisingly rough, windy, and crackly, making the tiny, rocking boat seem all the more fragile upon the moving swells. "We're hearing snapping shrimp," nods Paul. It's a sound like electric sparks or radio interference from another star. He sits Buddha-like in big mirrored shades at the back by the motor, taking it all in. "Some days you'll hear nothing but the shrimp for hours." Waves drum against the rubber boat, and the hydrophone cable brushes against algae and rocks way down below.

We get used to the background crackle and then it starts. First it is faint, but soon unmistakable. Far away, in the midst of the sonic mess, a faint howl begins: the rise, the fall, the squeal, the sweep. Out there in the sea somewhere, maybe five miles away, it's the song of a humpback whale, a music unlike any other.

The range is tremendous, from the bowed bass beats of a giant subsurface fiddle to the feedback squelches of a para-electric guitar. Each note is solid, emphatic, determined, beaming with feeling. It is one of the greatest sounds in the animal world, and very few people had heard it until the end of the 1960s.

This is quite surprising, because if you dive underwater in a place where humpbacks gather to breed, it is easy to hear them with no equipment at all. But no one seems to have thought to listen until the first recordings were made. There are only scattered references in old maritime literature to humpbacks making sounds we could hear above the surface. In 1856 Charles Nordhoff wrote, "a whale would sometimes get under the boat and there utter the most doleful groans, interspersed with a gurgling sound such as a drowning man might make. The first time I heard these sounds it was almost incomprehensible to me that they could proceed from a whale." In 1889 H. L. Aldrich wrote that "it has been known for a long while that humpback-whales . . . have their own peculiar cry, or as whalemen express it, 'sing.'" If whalers all knew this, they sure didn't say much about it. After small-boat whaling faded away, no one saw fit to mention the sound for nearly a century.

Scholars say these songs might be the source of the myth of the sirens beckoning to Odysseus as he lay down on the benches of his ship while drifting home, dreaming of Penelope. Until fairly recent times, whales were mostly seen as sources for the oil needed to fuel our society, that is, until petroleum was discovered to be a much more accessible source of the stuff. Melville in *Moby Dick* mused about what tones would come out if such great beasts could actually speak! Well, it turns out some species sing. Others cry, and still others only click.

When the human world first heard these sounds, our sense of whales suddenly changed. It is the song of the humpback whale that made us take notice and begin to care about these animals—the largest that have ever lived on Earth. We would never have been inspired to try and save the whale without being touched by its song. Greenpeace would never have woken the world up to the cruelties of whaling, and perhaps the whole ecology movement would never have gotten so strong. Humpback whale song is so emblematic of the early days of the green movement that some people laugh at me when I mention it, shelving the music of whales with past lives, weekend shamans, suburban yogis, and other fragments of our yearning for deeper meaning. All this is easier to dismiss as woo woo than take seriously.

Scientists who study humpback whale songs usually collect the music as data, take it home to the lab, and then pore over their recordings as computer transcriptions, discovering form, rhyme, change, and rhythm. They've been doing this for nearly forty years, yet humans still don't know what these songs are for or even how the whales produce them. They don't seem to fit the normal patterns of animal behavior.

There are people all over the world taking whale songs apart in the laboratory, but only Paul goes out every day just to listen. And he never wants to get too near the whales he hears. "Sure, I've heard a lot of close-up recordings, where people are chasing the whales in big fast boats. Yeah, I've heard those songs. Those whales sound pretty stressed out to me." He radiates a certain calm contentment with his listening life on the beach—with no sense of edge or doubt whatsoever.

No one who hears these darkly beautiful tones for themselves can easily forget them. Upon hearing the great song for the first time, whale pioneer Roger Payne said he heard the size of the ocean, "as if I had walked into a dark cave to hear wave after wave of echoes cascading back from the darkness beyond. . . . That's what whales do, give the ocean its voice." Most work on the meaning of these tones is far more prosaic, involving pages of calculations, summary charts that have a hard time containing the original beauty.

But Knapp doesn't care why whales sing, he's just out there on the water every day, listening. He has no money. And he has no job, except to take passing tourists out to listen to whales. Paul's work takes place on that raft. He brings as many people along with him as will come. You may pay as much or as little as you wish. He gets by on whatever his passengers offer him.

Knapp first heard these sounds twenty years ago, and he's been coming back to Brewer's Bay every year since. He's released two albums of humpback whale songs, and they are among the most sensitive and delicate recordings available. It's easy to see why. Knapp takes his time, waits for the whales, and knows that most of the time the sound will not be perfect—a good reason to go out again.

Out on the boat we hear overlapping, faint howls. Group singing, where several males make sounds that follow and interact with each other in a repeating mix of rhythms. To Knapp these simpler, chorusing whoops have an even more universal appeal than the long, solo songs. Blending with the high percussion of the snapping shrimp and the slow beat of the waves and swells, it's oceanic trance music.

We go back to shore to see if anyone else wants to come out. It's high season on Tortola and there's always someone wandering by. First we spy two women reading thick books on the beach. "Hey," Paul asks, "want to hear some whales?" Already nearly sunburned, they're ready to go.

Turns out they are ringers, biologists from Maine, who know plenty about whales and marine science. Still, they've never heard humpbacks in person, so they're ready to go. Back out at the edge of the bay the big beasts do not disappoint; the choir of overlapping voices is back. Their song blares through the speakers, rising out from water into air. I mention a curious fact I had recently read: since it is only the male humpbacks who sing, science would like to believe the song has something to do with attracting females. But no one has ever observed a female humpback approach a singing male.

"Well, biology has been dominated by men for more than a century," says one woman to the other, and her friend nods knowingly back. "They always come up with the same, simple explanation for everything." Seems like these two may have dealt with this issue before. "There has to be more to the story, and it's going to take a lot of long, hard listening to figure it out."

When it comes to song, scientists have spent a lot more time studying birds than whales, and the standard explanation is that males advertise how fit they are by singing long, complex songs. The females tended to choose certain kinds of songs over the generations and these are the songs that have prevailed. There is enough evidence to suggest this is often true, but it says very little about *what* the birds are singing, why it is often so musical. Science tends to say the particular preference of females is arbitrary, something like the whims of fashion.

If you speed up a humpback whale song it sounds just like a bird. It has the tonality of a catbird, perhaps, with the rhythmic precision of a nightingale—beats, quips, melodies put together in a precise, definitive way. Why should these musical principles appear in nature at such different scales? Maybe music is a part of nature itself, something evolution has produced on different lines, converging into some living beauty that whales, birds, and even humans can know. Music is much easier to appreciate than language, of course. You need not understand it in order to love it.

Our next guest, on the second trip of the day, is a computer programmer from The Hague. A ruffled figure camping alone by the beach, he's dressed in heavy clothes even though the weather is tropical. First off he is surprised that anyone would just come up and talk to him while he sat outside his tent. He seems genuinely touched by the invitation, having no idea there were whales anywhere near.

The sea has gotten a bit rougher and the raft is tiny, so the swells start to get to him and he blows his lunch over the starboard side. His face has turned green. But when he puts on the headphones a more pleasant color returns. He listens intently as if taking notes in his head. "Vel, surely zhis can be figured out, no? Zhere is clearly information content, ja? A machine could certainly help us grab zhe patterns."

So far human listeners have been better at making sense out of the logic of whale sounds than any automatic attempts at song recognition, though some believe machine learning and AI may soon change that. The U.S. Navy, who knew about humpback whale songs two decades before the general public, has published guidebooks to help sonar cadets distinguish whale noises from all of the underwater signals they are supposed to keep track of, such as those from enemy submarines and torpedoes. Whale sounds are just classified as "biologicals," nothing to get too worked up about.

After two runs I'm starting to get into this lifestyle; I can see why it appeals to Paul. All who hear the whales slow down, stop speaking for some minutes, and wonder in their own way about what messages may be out there.

"The whales are the artists," says Knapp, "I'm just trying to capture what they do. People need to hear this. I think it's good for us—to have a different vision. That the ocean has a singer, a spokesman, helps us to see the Earth as something else than we often do: a place with other life, other presence, other beauty. We are not the only musicians on this planet."

Back on shore the beach is getting warmer, and busier. Our third group consists of two forty-year-old tanned surfer babes and one teenage boy—boyfriend or son, I'm not sure. One of these women has been out before, and she just has to bring her friend, who gets real excited as the music begins. "Oh, listen to him, he's getting there, Geez Louise, all right, that's a good one, yeah baby! *Whoooee!*"

"What do you think the song is about?" I ask her from behind my wraparound sunglasses.

Staring at me with disbelief she says, "I thought that was obvious. It's got to be something to do with mating. He really sounded into it."

She could get a job as a whale scientist.

This is my first time out to hear whales, and I've learned in just one day how these humpback songs hold countless meanings for those of us lucky enough to hear them. Paul reckons he's taken more than a thousand people out, no more than three at a time, to hear this music of the sea. "I find something truthful and relevant in the song. If I didn't, I wouldn't spend my time doing this. Those are two big words, truthful and relevant. If you find that you've got something worth pursuing."

In all my travels playing music with animals, whether they are nightingales, cicadas, coyotes, or water scorpions, I'm most interested in what kind of music we can make together. Without knowing what the animals think about such interaction, some kind of interspecies music is made. Either art spans its way into the more-than-human world, or it's all some kind of ridiculous stunt. Often I'm not at all sure. Making music with the great singers of the sea has been my fondest adventure.

Of course there are many obstacles to such a dream. Humans and whales live in quite different worlds. Whales travel far under water and can barely see. Sound is everything to them. It tells them much more about their environment than what we hear. Clicks and pulses help some species navigate by how the sound goes out and back—echolocation, or finding your way through sound. Humans have turned such ability into technology and developed sonar. But whales can still do it better than us. A beluga can detect the precise shape of an object under water behind an opaque screen. No human machine can do that yet.

Thousands of meters way down in the ocean, there is a realm known as the deep sound channel, where the lowest tones—the subsonic groans made by blue whales and the pulses made by fin whales far beneath the range of human hearing—may be able to carry for hundreds and even thousands of miles. We know it is feasible for a distant whale to hear the lowest tones of her possible mate so far away, but we have no idea if she actually listens. A thousand mile song—the scale of it defies comprehension. Great whales, with their huge brains and slow metabolism, have an experience of life nearly impossible to fathom. After centuries of thinking of whales as little more than oil, meat, and blubber, we now see them as icons for the saving and cherishing of the Earth, reminders that nature will always remain more than we can ever understand.

How did the whale rise in our consciousness in this whole new way? It all began with the sound. Hearing the humpback whale song from a rubber raft splashing in the Virgin Islands waves is unforgettable, but many more people have been touched by the recording *Songs of the Humpback Whale*, released by the Wildlife Conservation Society at the Bronx Zoo in 1970, later included as a "sound page" in an issue of *National Geographic* in 1979. They printed ten million of those and sent them around the world; it was the largest single pressing of any record in history, a record that is likely never to be broken. In 2021 *National Geographic* considered whale song as music in a new way, with my bass clarinet overlayed upon it. I'll tell that whole story in the new final chapter of this book.

Once you hear the music, it is hard to think of killing any animal that can sing so beautifully. True, most other species make less musical noises, but they are far from silent. Sperm whales have elaborate clicking patterns that some have likened to West African polyrhythms. Killer whales, also called orcas, have a wide range of whistles and slaps that vary tremendously from pod to pod. Smaller, white beluga whales have the widest range of pitches from low to high, and textures from shriek to creak to ratchet to noise. That species has been admired long enough by sailors to earn the nickname of "sea canary." But humpbacks have the most extended and resonant tones and phrases, and it is they who have moved humans with the greatest music in the natural world.

It's a song that never stays the same. Each winter the Atlantic humpback whales return to the sheltered waters of the Caribbean, where the males sing a song that continuously changes through the course of the breeding season. No other animal changes its song so rapidly in a single season, and we have no idea why or how the whales do this. Do they follow the latest hit song just because it's popular, or is there a message hidden in their variations? We don't know, and we hardly know what it would take to find out.

The world of whale sounds reverberates deep under the sea, from hundreds of meters to hundreds of miles. It's completely alien to the soundscape of humanity, shouting and singing up in the air, where songs and speech can't dream of carrying so far. Under the surface lie deep booming patterns, and perhaps approaching it as music can help us make sense of it where words and logic fail.

As I did with birds, I have played my clarinet live with whales for many years now and have learned something of how they might react. Have I had a hand in the changing of their song? Or have I been just another strange sound to challenge or ignore? On this trip with Paul, my first, I was only listening; but later I'll head to Canada for orcas, to Russia for belugas, and many times to Hawaii and the Dominican Republic for humpbacks. I'm prepared to join in wherever I am.

How can a human play music with a whale? I have to cross the sound barrier into their world. Playing my clarinet on the shore or on a boat, I get a microphone that picks up my sound and plug it into an underwater speaker that can broadcast these melodies way down into the world of the whales. Mine is a watery clarinet sound that is also able to travel at least a few hundred meters into the sea. A hydrophone is positioned not far away, and that will bring the mix of human and whale sound back together as I listen to it through headphones above the surface.

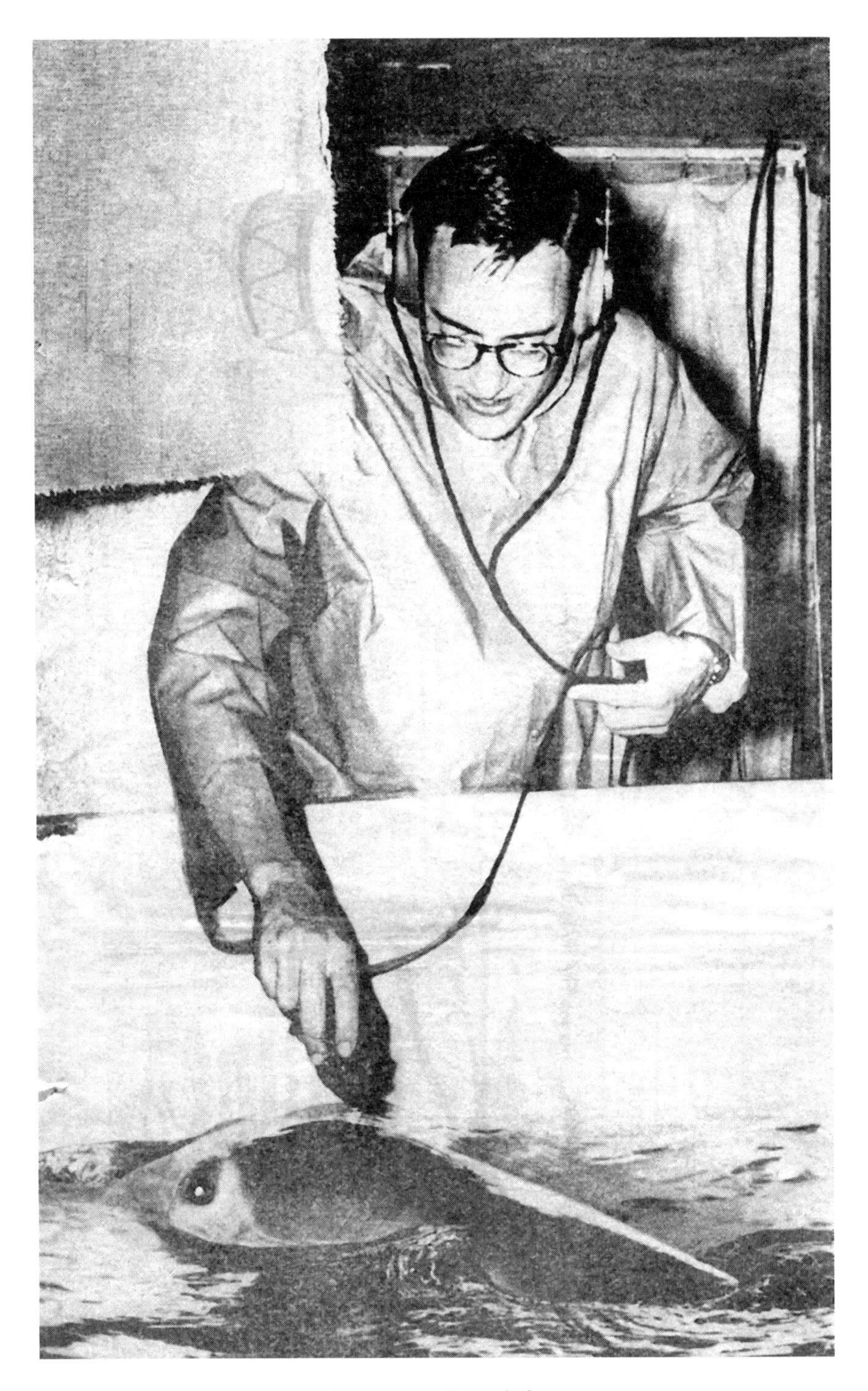

*Fig. 1. Scott McVay and Elvar*

Why would a whale care about human sound? There is a history of experiments demonstrating that dolphins, belugas, and even orcas are curious and enjoy playing around with sound. The larger the whale, the less interested they seem to be in our tunes. No one has successfully gotten a humpback whale to interact with human music—that at least is the official line. I hope to discover otherwise. Perhaps humpbacks might be inspired to heed our music as we have learned to care about theirs.

If we imagine animal sounds to be a kind of language, we will want to decipher them, to translate every nuance into a functional message. It is hard to correlate specific whale behaviors with specific noises. Some sounds help them navigate, some help them say hello. Most probably serve to hold groups of them together as part of ritual or culture—the way of being a whale. They live and sense through sounds both complex and beautiful. Consider them as music, and then we may enjoy them before explaining every noise away.

Human description cannot contain such reality, such meaning lies beyond us. The sounds are so iconic they have become ironic, the prototypical New Age shtick, just another excess of the 1970s. In this age where music continues to open up to the possibilities of just about any sound, it's time to bring whale song back into the public ear. Especially now that the internet makes any music endlessly available, all the time. In this world of such riches we need to widen our sense of what to listen to and how to listen with the greatest possible ears.

Nothing we know about whales is certain. Are they intelligent? Can they hear each other from afar? Are they as interested in us as we are in them? Can they tell a good song from a bad one? We are still at the very early stage of figuring out these animals and learning to appreciate them. Hopefully we and the whales will both be around long enough to let the relationship grow. To find better answers we need to ask better questions, and more people need to listen to this undersea music. More people need to listen, and to care.

Most of us are made to read *Moby Dick* sometime in college. We're usually taught that it's a great adventure story about obsession, mad leadership gone awry, or the hopelessness of extreme human quests to dominate nature. By the time it became popular, fifty years after its author's death, it was a window into an entire industry that was no longer familiar to most who picked up the book.

Scott McVay was luckier than most of us. At Princeton in the 1950s, he took a course with Lawrence Thompson, one of the top Melville scholars of the day. "Don't skip the whale stuff," said Thompson. Little did he know this advice would lead McVay to spend a lifetime devoted to whales. McVay, along with Roger Payne, is the man who discovered that the song of the humpback whale is a true composition: regular, rhythmic, full of patterns and organization.

McVay is one of those cheerful, well-connected people who remembers everyone he's ever met. Tall, in his late-eighties with an enthusiastic grin, he usually sports a bowtie. Conservationist and poet, he has worked as director of some of the largest and most influential philanthropic foundations in America. His greatest passion has been for whales, and his essential role in the effort to save them has been overlooked in favor of more public, outgoing figures. He's a behind-the-scenes guy, eager to put on events and make things happen. McVay is the key man in the story of how the whale came to be saved by its beautiful song. With a glint in his eye, he tells me his story.

As a young administrator at Princeton University, his alma mater, he happened to attend a lecture by maverick cetacean scientist John Lilly. It was 1961, and with the recent success of his book *Man and Dolphin*, Lilly was well on his way to becoming one of the most famous popular-science writers of the day. The public was becoming transfixed by the possibility that these underwater creatures possessed great but alien intelligence. Sure, many people were impressed by Lilly's ideas, but only Scott McVay typed out eighty-three questions about the book to hand to the author after his speech. They rode the train together back to New York, and Lilly invited to work for him.

Scott had two young kids and a fine job as the university's first recording secretary, with fifteen people under him. He wasn't ready to take off. But Lilly knew a good man when he met one and kept hounding him. Two years later McVay took his family down to Coconut Grove, Florida and stayed there for two years. He was assigned to work with the most precocious dolphin, an eager learner named Elvar.

There are basically two types of whales. First are the mysticetes, also known as baleen whales, who comb through the water with huge mouths agape stuffing all kinds of tiny creatures into their gullets. This family includes most of the biggest species: blues, fins, humpbacks, minkes, seis, rights, and bowheads. Next are the odonticetes, or toothed whales, the largest of which is the sperm whale, the Earth's biggest carnivore, who eats mainly squid. This group includes killer whales or orcas, belugas, pilots, and all species of porpoise and dolphins (who are basically just small whales). Since they are much more manageable in captivity, most of the controlled experiments on cetacean hearing and learning have been done on dolphins. They were the main subject animals in Lilly's research.

McVay's task was to teach Elvar the dolphin a set of 198 vowel-consonant combinations, through regular attempts at interspecies interaction. McVay would say "*ease ooze oar*" and the animal would come back with something like "*ea oo awr*" and they'd try this over and over again. Elvar couldn't really hear the lower frequencies, so McVay sped up his voice four times on tape, and then it began to sound like what Lilly called "delphinese," the supposed language of the dolphins.

But all the while, McVay felt they were approaching the problem from the wrong side: "I did exactly what Lilly wanted done, but after a while it seemed like we were brainwashing the dolphin with regard to the English language. . . . I became interested in kind of the opposite, listening to what the dolphins themselves were doing, to try to figure out *their* sounds, not to teach them ours."

He left the Lilly lab and returned to Princeton, where he became assistant to the president of the university. Yet McVay's curiosity about dolphins and whales, and what our proper human relationship to them should be, had only increased. He was haunted by a line from *Moby Dick* on the uncertainty of the whale's future, which Melville already suspected in 1851: "The moot point is, whether Leviathan can long endure so wide a chase, and so remorseless a havoc; whether he must at last be exterminated from the waters."

Back in the nineteenth century, Melville answered with a resounding "no!" But that was long before Norwegian whaler Svend Foyn's invention of the explosive cannon harpoon around 1890 and the transformation of whaling vessels from lone, struggling ships into the factories of capture, killing, and processing they became in the twentieth century. By the end of World War II the populations of most whale species had been so depleted by hunting that the International Whaling Commission was formed to try to regulate whale "stocks" so that the industry could continue in some sustainable fashion. But the killing continued.

In 1965 the commission conducted the first extensive survey on the number of whales remaining. The populations of blues and humpbacks had plummeted to near zero, and most other commercially hunted species were in trouble as well. Perhaps the whale fishery as a whole was an unsustainable enterprise whose time was now gone. If whaling as an industry was to survive, the hunting nations would have to impose careful quotas on the number of animals they slaughtered.

The general public really gave the matter no thought whatsoever, even if they had all read *Moby Dick*. Scientists realized the word needed to get out, so they encouraged McVay, an excellent writer and communicator, to write an article for *Scientific American* in the summer of 1966, entitled "The Last of the Great Whales." The piece included very clear and alarming graphs revealing how only six decades of intense, industrialized whaling in the Antarctic had decimated whale numbers, urging the Commission and the global community to reconsider whether whaling as an industry was obsolete, since so few animals were left to kill. "What troubles me," wrote McVay, "is the prospect that the whales, possessing the largest brains on earth and gloriously unique in the scheme of living things, will be gone from the earth before we may be able to understand them."

The ideas of ecology were just beginning to be taken seriously in the public air. Commercial whaling began to be seen as a symbol of unstoppable human greed. The ideas of ecology were just beginning to seep into the general mood. John Lilly's dolphins were inspiring millions. Television gave us Flipper, Sea World had Shamu, and the film version of *The Day of the Dolphin* was already in the works. People fell in love with dolphins and whales because of the way the media offered them to us: as sensitive, even telepathic creatures who showed compassion for our kind, aiding drowning swimmers and helping fishermen with their catch. Meanwhile, a few privileged individuals were hearing some very strange underwater sounds. . . .

Frank Watlington had a very secret job. The Navy stationed him in Bermuda at the dawn of the Cold War to listen under water. Already by the early 1950s, the Palisades Sonar Station at St. David's, Bermuda, had primitive hydrophones rigged seven hundred meters down in the ocean so that clear recordings could be made without too much surface noise from traveling ships. Watlington's main task was to listen for Russian submarines that could have been anywhere, at any time, threatening our freedom with no mercy and no bounds.

Instead of picking up submarines the hydrophones called up all manner of strange, mysterious sounds: "We had become quite familiar with the sounds from propeller-driven ships passing the islands, with explosive signals, and with sounds from West Indian seismic activity, but this new sound was quite unfamiliar, coming in with breathtaking clarity." Watlington was determined to discover the source, so he went back to his laboratory to build better hydrophones that could more effectively pick up the strange sounds. A few years later a better system was installed and attached to a buoy in fairly shallow water.

It was not until 1955 that Watlington was able to sight three whales within fifty feet of the hydrophone, and then the sounds came in clearer than ever before or since, like Paul Knapp's magic Valentine's Day recording. Now he knew that humpback whales were the source of this marvelous noise. "Such success was pure luck, as I have discovered many times since." Further listening revealed that only humpbacks made anything like this complex music.

For fifteen years, no one besides Watlington and a few Woods Hole acousticians knew the beauty and range of humpback whale sounds. And no one who was privy to this classified information spent much time listening to the songs—they weren't coming from the enemy, so why bother? Plus, Watlington was worried the whaling industry might find out about these sounds and gain yet another weapon in their battle to exterminate the last of the great undersea beasts. If whalers could track whales by their songs, they might be able to kill them even faster.

In 1967 Watlington handed his tapes over to Roger Payne, the first person he thought could be trusted with them. Payne was beginning to make a name for himself as the great public voice of whale science and whale conservation around the world. He didn't really have the time or the inclination to study the tapes in depth either, and he knew Scott McVay had worked on dolphin vocalizations with Lilly. So he passed the tapes along to McVay.

Princeton University had one of the few sonograph machines capable of printing out the details of the sound in a form that could help make sense of its organization. This device was originally invented during World War II as a possible aid to help the deaf learn to speak, but instead it was put to new uses because of its ability to accurately visualize complex sounds otherwise hard for humans to describe. The machine had already helped scientists decipher bird sounds, and Payne thought the sonograph might shed some light on the strange wails of the humpback whale.

Today any personal computer running programs such as Amadeus, Audacity, Audition, iZotope RX, or Raven can print out in minutes a sonogram of a sound of any length; but in those days it was quite an effort to get the machine to draw what it was hearing. Mark Konishi, one of the pioneer birdsong scientists, had a lab in Eno Hall at Princeton and McVay would work there late nights and weekends, painstakingly printing out one syllable of the whale song at a time and then taping the pages together into a giant scroll. Humpback whale songs run uninterrupted for up to thirty minutes at a time, so this was a lot more time consuming than working with the song of your average fast-talking songbird.

"What is science if not bean-counting and patient observation?" McVay smiles. "I literally spread these printouts on the living room floor and kept looking and looking at them. My wife Hella, a math teacher, came over and looked along with me. We finally looked at each other in astonishment and said, 'My God, it repeats.'" There was a regular pattern with form and shape. With the alien song all graphed out, frequency against time, the McVays could see its intelligible structure.

Whereas birds perceive information about twice as fast as we do, whales must do so much slower for them to follow such drawn-out patterns. Speed the whale up or slow down the bird, and they begin to show the same senses of organization at work at different levels. What does this prove? Natural selection may use randomness to advance, but its methods somehow generate beauty.

"Hella and I looked at each other and had the same thought: we've got to go to Roger with this, because no one is going to believe us." McVay has always known how to pick the right allies. When Payne saw what the McVays had uncovered in the structure of the song, his jaw dropped and he calmly announced, "Maybe we should publish this together. Who should be the lead name on this?"

"Let it be you," said McVay, with honest modesty. "You're a card-carrying biologist, this is your life's work. I'm just an English major who loved *Moby Dick.*"

"Songs of Humpback Whales," by Roger S. Payne and Scott McVay, appeared in the influential journal *Science* in 1971. The paper was accepted more than a year in advance of publication, but they wanted it to appear on the cover, and all the earlier covers were taken. It was worth the wait because this is one of the most beautiful covers the journal ever published, carefully depicting the beauty and structure of the song visible in one glance. It is still the best and clearest paper ever written on whale song, with carefully arranged tracings of the early sonogram printouts and a real attempt at structural, almost musical analysis of the kind rarely attempted in bioacoustics. The illustrations show the enigmatic but nearly hieroglyphic structure of the repeating whale syllables, like markings from another planet or civilization.

This was the humpback whale's big break. Until then, these animals had been hunted, feared, admired, but not really known. Suddenly they were capable of something few animals could boast of: a long, developed, extended vocal performance that looked like it contained a whole range of information. And it was something that could be genuinely discovered and appreciated by every new person that heard it. The paper contained a most remarkable sentence: The humpback whales "emit a series of *surprisingly beautiful sounds.*" More than fifty years have gone by, and no other scientific paper about whale songs has dared to mention beauty. Payne and McVay remain pioneers in forging a bond between art and science. In space we had reached the moon, and in the oceanic depths we had heard whales sing.

The record *Songs of the Humpback Whale* came out first, in the spring of 1970. It was pressed by an outfit called Communications Research Machines in Del Mar, California, who mostly published psychology books. It came in a most unusual jacket: all white, with a painting of a majestic humpback leaping out of the water next to a small rowboat glued onto the front. No words, no title, nothing on the back at all. Inside is a full-size, thirty-six-page booklet with texts in English, Russian and Japanese, urging us to save the whale. "Listen to him singing far below the turmoil and ceaseless motion of the surface. . . . From that profoundly peaceful place a voice calls us to Turn Back." There's a chart showing just how much bigger whales can get than dinosaurs. There are sonograms of the structure of the whale songs and detailed charts on the decimation of whale populations thanks to human hunting. The inside back cover has a photo of bloody guts on the deck of whaling ship. Listeners were urged to do whatever we could to halt the senseless killing of these great, musical animals. On the edge of the circular label on the record itself, these words appear over and over again until they loop upon themselves: "turn back turn back turn back turn back. . . ."

The record sleeve contains detailed instructions on how to listen to the whale music. As if presaging the future personal soundscapes of Walkmans and iPods, the sleeve urges us to put on headphones and surround ourselves with the noisy ocean full of sound. With your ears covered, the sonic ecosystem takes over your senses. Postcards in the back of the booklet could be used to order more copies at $9.95, with the money going to the Whale Campaign at the New York Zoological Society, the Bronx Zoo.

The first review appeared in *Time* magazine, by a writer obviously flummoxed by an inability to describe what he was hearing, music that "might have come from the throat of a 40-ton canary to the rumble of a stupendous Model T with a cracked muffler." Albert Goldman in *Life* likened the sound of the water burbling past the hydrophones to "the waters in Wagner's Rhine, while down in the depths you hear squeegee squeals and two-cylinder *brrrrrrps* and other-worldly cries that arc across acoustic space like particles in a cloud chamber or the inconsequent anguish of an atonal violin." Whale song sounded of the moment, totally au courant.

It was Jon Carroll at *Rolling Stone* who really got it, realizing that this song was destined to be a hit: "This is a good record, dig? Not just an interesting record, or, God help us, a *trippy* record. . . . It's especially good for late at night and peaceful, together moments. It stretches your mind to encompass alien art forms. The kids can't dance to it, but they can't dance to Satie either." Carroll recognized that these humpbacks sounded like the electronic music that was just starting to get beamed into our consciousness, but he preferred the whales, because they were "without the sterile, hyper-intelligence trip that much electronic music gets into." (I asked Carroll, decades later a columnist for the *San Francisco Chronicle*, whether he remembered the first moment he ever heard a whale song, and he said, "No. We were all doing a lot of drugs in those days.")

Whale song was thus anointed as hip, serious, and 100 percent natural. The music came out of the blue and people loved it like nothing else. In February 1971 Joseph Morgenstern in *Newsweek* reported that 45,000 copies had been sold, and the record was reissued as a regular disk without the pages of explanation and conservation appeal. Morgenstern was a bit concerned, "I must confess to some dismay when I learned that the kids were lacing their whale songs with pot. It seemed as if the counterculture were co-opting a straight sense of wonder."

It is rare for humanity to come across a new experience that is impossible to expect or describe. No one imagined great whales could make such great sounds, and this was an age when new experience was especially treasured, sought out, and blended with a rush of sensory possibilities. We dreamed of a better, more joyful world, and singing whales would be part of it.

The release of the record changed the lives of Roger Payne and Scott McVay. "The impact of the song," said McVay, "was huge and staggering." The two men appeared on *Good Morning America*, and McVay alone on the *Arthur Godfrey Show*, where he was asked, "Scott, tell me what an endangered species is."

"You sir," said McVay, "are an endangered species."

After taping a whole hour on whale song, Godfrey said, "Look, we like to tape three of these programs at a time. Today is Valentine's Day. What do you know about their courtship and mating?"

"Enough to go on for another hour."

Seizing the moment, McVay embarked upon a remarkable journey to Japan to try to get this major whaling country to take a second look at their whale-killing ways. Armed with a dozen copies of *Songs of the Humpback Whale*, with its Japanese liner notes, he took off on August 6, 1970.

In advance of the journey, a committee of six eminent Japanese scientists was created to tackle the issue of whale conservation in Japan, under the leadership of Seiji Kaya, former president of Tokyo University. The group included the former head of Japan's atomic energy commission, the director of their national institute of genetics, and the director of Japan's primate research center. Other allies included the young novelist Kenzaburo Oe, who later went on to win the Nobel Prize in literature; Kunio Maekawa, designer of the Japanese pavilion at the New York World's Fair; and the composer Toru Takemitsu.

McVay appeared on the biggest Japanese radio and television shows, charming millions of listeners and viewers with the record of the songs and his paeans to their beauty. "Imagine the day," he regaled them, through a translator, "when you can take your kids on a photographic safari underwater to follow a pod of whales. What splendid sport it would be to see a whale moving through its water world, to film that great graceful hulk in the well-illuminated waters near the surface as readily as you would an elephant on the Serengeti Plain." He paused for emphasis. "Yet as certainly as technology will catch up with this dream, so do we seem as certainly bent on cancelling that chance by the almost unchecked slaughter of the remaining whales. Today, every twelve minutes the charge in the head of a harpoon explodes in a whale's body."

The Japanese public was touched by the humpback whale song in a manner far deeper than the Americans, perhaps because whaling was much more relevant to their culture. The popular science fiction writer Sakyo Komatsu said, "You have opened our eyes and minds to a new frontier for the human soul." McVay graciously offered the Japanese gentle questions and koans on these beautiful animals of the sea. He quoted Thoreau: "Can he who has discovered only the values of whale bone and whale oil be said to have discovered the true use of the whale?" When he met Kota Hoketsu, chairman of the board of one of the biggest Japanese whaling concerns, McVay played him the song. Hoketsu shook his head solemnly and whispered, "*We didn't know, we didn't know.*" He pledged to play the song at the company's next board meeting.

Arrangements were made with the Asahi Company to release a mass-market edition of the whale song record as a flexible plastic "sound page," which could be printed in book form. It was to sell for the equivalent of

$3, one third the price of the American edition, so as to be accessible to the largest possible number of people. Kenzaburo Oe published a moving story, "The Day the Whale Becomes Extinct," where an old man announces that the last whale has died and a young child asks the writer, "What on earth was a whale?"

> I tried to explain to him about whales, but since I didn't remember a thing about whales myself, I clammed up. Then I realized with a sense of fathomless annihilation that the old man of just now was the "me" of the past and the child none other than the "me" of the future. The face of the child, looking up silently at me, distinct in the evening light, was crumbling away.

In 1970 the fate of the whale lay decisively in human (and especially Japanese) hands. The nation was moved to tears as whale song resounded in everyone's ears. It is nearly forty years later, and Japan *still* wants to kill whales, defying the International Whaling Commission's 1986 ban on commercial whaling, even though most of that country's citizens are against it.

Upon his return, McVay asked his good friend Mark Konishi how the Japanese could seem to love the whale songs so much and still want to hunt them until they're gone. "In Japan," smiled Konishi, "ethics . . . *very slippery*."

As they may be everywhere. Hoketsu, captain of the whaling industry, was also famous for being a bird conservationist. "It is very simple," he gestured to McVay over dinner. "With my left hand I stroke the birds and with my right, *shaft* the whales."

McVay was moved enough by his journey to write this haiku-like poem entitled "The Alternative":

> If we do nothing,
> in a few years
> the whale will live only
> as a legend,
> a marine Daidarabochi who,
> it was said,
> sang unspeakably beautiful songs.

(A *Daidarabochi* is a mythical giant, familiar to some from the Miyazaki film *Princess Mononoke* as "Didarabocchi.") Later this elegant message appeared on a save-the-whales poster from the Ocean Conservancy, above a beautiful whale image, only the author's name was listed as "the distinguished and frail poet Osaka."

McVay laughs. "I wrote that *in* Osaka. Don't know where they got the 'frail' or 'distinguished' part. On the next printing they put my name back in, but that takes the whiz out of the story! If I ever publish a book of my poems, I'll entitle it *Whales Sing*." (That book came out in 2013, followed two years later by the memoir *Surprise Encounters with Artists and Scientists, Whales and Other Living Things*.)

McVay performed the title poem from his future book on the occasion of his retirement from his position as director of the Geraldine R. Dodge Foundation in 1998, at the biannual poetry festival the foundation supports, the largest such event in the country. The Paul Winter Consort, the musical group most known for its integration of whale and human melodies, accompanied him:

> If the whale's voice be stilled,
> then I shall be quieted utterly.
> I have this tape
> and I'll listen
> until I know each linked sound and
> can play them all in
> my mind whenever. . . .
>
> Before Watlington we intuited
> such miracles—
> now knowledge almost unbearable
> saturates the senses,
> charges the mind.
>
> Music must be the language
> that blends
> two otherwise separate paths
> for a way,

making each journey more
than either alone,
linking all journeys.

A noble song fills the head
—implications shake me.
Share first with one or two
well chosen
and then with all who care.

As I heard this great activist intone the tale of the cause closest to his heart, I felt ready to follow my own quest for the "grail" of the whale's song. McVay is one man who realizes art is more than a metaphor when it comes to saving nature. There has always been nature in poetry, but here is a man who discovered a poetry in nature when he first identified the stanzas and rhymes in whales' songs. In his 1971 article in *Natural History*, "Can Leviathan Endure So Wide a Chase?" McVay sums it all up once more: "To leave the oceans barren of whales is as unthinkable as taking all music away. . . ."

Going through many editions before migrating to the internet like all other recorded music, the record *Songs of the Humpback Whale* became the best-selling nature recording of all time, a multiplatinum album that has sold millions of copies, right up there with Pink Floyd's *The Wall* and Paul Simon's *Graceland*. Without the effect of that whale song record on the millions of people who heard it, there might never have arisen a movement to save the whales, transforming their image from oil and blubber to gentle serenaders of the sea—turning the majority of the world's people against whaling, probably forever. In 2015 I offered up an album called *New Songs of the Humpback Whale*, to present the finest recordings made by many scientists in recent years, to let the listening public decide if our more precise, digital technology sounds any better than those first wondrous Watlington tapes. I say more about this endeavor in the final chapter of this book.

After fifty years of conservation efforts by McVay and thousands of fellow activists, whale music can still be heard. Back on Tortola, Paul Knapp still goes out every day to make sure, fifteen years after I last saw him there. "It's good for me to go out every couple hours to hear what's happening out there. Things do change, and you have to keep up with it, things are happening, and if you are not aware of it you're not going to be able to capture it on tape." Good thing there are witnesses on the water taking in all the changes—someone has to notice such things to remind us what is truly worth saving.

# 2

## GONNA GROW FINS

## Humans Take Up with Whale Music

On January 3, 1971, New York Times critic Harold Schonberg summed up the previous year's highlights. In music, 1970 was surely the year of the whale. Luminaries from classical, pop, and folk music were so turned on by this amazing sound that they began to integrate it into their own work up here in the human world, safe on dry land. Judy Collins had a hit song with "Farewell to Tarwathie," and Pete Seeger wrote a rousing tune in a minor key, "The Song of the World's Last Whale." The first classical work to make use of whale sounds was Alan Hovhaness's "And God Created Great Whales."

After Roger Payne gave his tapes to Scott McVay for scientific analysis, he also thought a qualified composer should have access to them. New York Philharmonic conductor André Kostelanetz suggested Alan Hovhaness, a populist composer whose works blend his Armenian heritage with a cinematic response to the rhythms and moods of nature. The composer started work on "And God Created Great Whales" before the record *Songs of the Humpback Whale* came out, and his piece premiered on June 13, 1970, around the time the disk was first released.

Kostelanetz probably chose Hovhaness because of the booming, Orientalist quality of his music, with its heavily orchestrated modal melodies. Hovhaness never went in for the experimental, atonal excesses of modern music, so it was a safe bet he would produce something audiences would warm to. Yet the introduction of whales led the composer to write his strangest, freest, and most famous work.

Hovhaness claimed he was writing a music of the dawn of time, long before humanity appeared on Earth. Usually his music makes use of grandiose, unison melodies with a solemn Middle Eastern feel.

But when he had to incorporate material as unearthly as whale songs, he made his orchestra do unconventional things: improvise around the soloing whale, screech up and down the strings, offer a trombone *blatt* or a bass fiddle *thunk*. He bent his method to work more the way nature works, following rules only loosely, with results that sometimes surprise. After the whale section stops, the music returns to his usual orchestral exotica.

Hovhaness played his idea to Kostelanetz, who was not impressed. "That idea is too Oriental," said the conductor. "The whales don't sing Oriental music."

"But they do," Hovhaness defended himself. "That's the whole thing, they really do!" He took a five-note theme from one of his early operas, and Kostelanetz was fine with that.

*Time* magazine found "the eerie whale songs" to be a "natural complement to the mystical music of Hovhaness." Their critic wasn't sure if the thundering final applause was for the whales or for the composer, who "beamed like an ecologist, announcing that we've got to preserve everything we can on this planet. It's God's own little spaceship. Everything counts." The *New York Times* critic Donal Henehan wasn't quite sure the different species' voices met on an even keel. "His whales spoke profoundly, but [the orchestra] stayed on the surface. . . . Faced with such an irresistible soloist, Mr. Hovhaness must have suspected he would be harpooned." Henehan was not impressed by the "commonplace black-key melody, conjuring up the sea by unmeasured bowing and overlapping patterns, setting brass and percussion to echo the real thing." Too obvious, too close. It was music imitating nature without thinking enough about how.

In the mid-1980s Hovhaness added some choral parts to the piece and a grander version was performed *for* three killer whales at the Vancouver Aquarium, who jumped and leaped along with the human performers. The whole thing was documented in Barbara Willis Sweete's film *Whalesong*.

Toward the end of his life Hovhaness used to say "And God Created Great Whales" was the one piece he regretted having written. I suspect that's because he realized the composed parts of his music were no match for the power of the whale's voice and that the ensembles he worked with were not sufficiently prepared to improvise the way they ought to. Yet it remains his most performed work, perhaps the first example of officially sanctioned interspecies music. Those few musicians who dare to cross species lines will at least be remembered for their willingness to try.

Shortly after *Songs of the Humpback Whale* came out, Scott McVay was to give a lecture at New York's austere, oak-paneled Explorers Club, a gathering place for adventurers for more than a century. Word got out that Judy Collins, one of the most popular singers of the day, was interested in attending. Only problem was, at that time the club did not allow women through its doors. McVay thought such a rule entirely ridiculous and suggested he do two talks, one for men and another for "the general public." With that idea, the club relented, and Ms. Collins, known for her pure and quivering voice, got to hear the song of the humpback whale for the first time.

Immediately she was inspired to work the whale song into the album she was recording, which would be entitled *Whales and Nightingales* when it came out in November 1970. The most stirring song on the record is a solo performance of Judy singing a Scottish whaling shanty, "Farewell to Tarwathie," accompanied only by a particularly hypnotic track from *Songs of the Humpback Whale*. The whale part has a little delay added to it, so a slight echoing rhythm appears at the beginning, giving the tune a very subtle electronic beat. Judy sings crisply and mournfully of the loneliness of the whale hunt off frozen Arctic coasts:

> Farewell to Tarwathie
> Adieu Mormond Hill
> And the dear land of Crimmond
> I bid you farewell
> I'm bound off for Greenland
> And ready to sail
> In hopes to find riches
> In hunting the whale. . . .
>
> The cold coast of Greenland
> Is barren and bare
> No seed time nor harvest
> Is ever known there
> And the birds here sing sweetly
> In mountain and dale
> But there's no bird in Greenland
> To sing to the whale

This fine killing song wafts over the whale's own lament. Toward the end of the piece Judy modulates up a step, and the whale backup sounds just as much in tune, revealing how this animal music comes from a whole different harmonic world. As the sad tune fades away, Judy and the whale disappear into a wash of reverb, rewarding the listener with the thoughtful echo of deep oceans, where whale and human songs wash together over distant leagues.

Without any explicit preachiness, this song is the purest and most moving of any of the first examples of human/whale music. It certainly touched a chord with the public, for the album went gold a few months after its release, selling more than half a million copies by the spring of 1971, making it the most successful human/cetacean collaboration in the popularity contest that is the music business.

As the number of whale-inspired songs increased, the numbers of real whales dwindled in the ocean depths as we humans kept killing them. I don't want you to be particularly impressed by all the money made off of whale songs in those days, but the point is that these songs were on the airwaves, and whale music was filtering into human consciousness on many levels. Brought to human ears through the latest technology, they were ancient and contemporary all at once.

Right at the moment when we believed in hope, peace, and revolution, the song of the humpback whale appeared in our midst. We wanted causes to believe in, and the whales needed our help, and were making all these sounds for who knows what reason. Why not hear it as a cry for aid, a chance to reach out, a push for humanity to find a better way to fit into the world?

Of course those whales are singing for each other, not for us. If we listen to them closely and take them seriously, we honor our place in the bigger scheme of things. As we strain to appreciate sounds that are larger than we are, perhaps we move a little closer to the greatness of the natural world—the most important thing we can bear witness to, as the thinkers, questioners, and chroniclers of nature. To truly be human we must listen well.

Not everyone liked what they heard. Some early reviews of *Songs of the Humpback Whale* recognized that these underwater melodies might have something in common with the experimental human music of the day. Paul Kresh in *Stereo Review* wrote that "the humpback whale sounds more like a mewling cat than a nightingale, with echoey electronic overtones that should prove no threat to Morton Subotnick or György Ligeti, although they may be of some help to the whale in locating his friends, if he has any."

I called up Subotnick, thirty-seven years after that review was written, and asked him about this. "Wow," he was surprised. "That's certainly an insult to the whales."

Back in the day, he didn't think the whales sounded anything like the electronic music of the late sixties, such as his own pioneering work "Silver Apples of the Moon," the first classical commission for synthesizer. He does remember what it was like to first hear the deep, profound music of another species: "It wasn't primal, it wasn't animal-like. It was beyond any musical instrument we could then imagine."

The whales sounded nothing like a sine tone or white noise (the basic building blocks of early electronic music) yet the newly found animal sounds reminded young composers and performers that there were unknown worlds of music still to explore. "We were desperate to define a new music for ourselves, and we were thrilled to discover such beautiful sounds were out there that had never even been considered before." Still, like most human composers, Subotnick was content to let whale song occupy some sacred place beyond: "It was an Ur-music, almost a religious experience."

Hovhaness was the first classical composer to have access to the prized Bermuda whale tapes, but George Crumb was sent the recordings only a few months later. Like most composers, Crumb may not be a household name, but his music is incredibly powerful, subtle, and unique, full of unusual sounds and deep silences made by unconventional techniques like singing into a flute or preparing a piano with paper clips and string to give it a brand new sound. This is music that reveals a genuine astonishment in the sheer beauty of sounds, nowhere as evident as in his delicate and subtle piece, "Vox Balaenae"—or "Voice of the Whale."

Crumb looks and sounds like the guy who's run your local hardware store for fifty years. He is not impressed by the ways some of his predecessors have integrated natural sound into their music: "Some of the real birds make Beethoven's Pastorale Symphony birds look like *pikers*. He had no sense of the sounds of nature. Now eighty years later with Debussy, things started to change. . . . Not now, Yoda, *quiet!*" says the great master to his dog, a nervous terrier.

We're speaking at his plain ranch house in the suburbs of Philadelphia. I ask Crumb, seventy-six years old at the time, if he thinks contemporary composers are closer to making good use of nature's sounds than their forebears. "Oh, certainly in the twentieth century, in Ives and Bartók I begin to hear nature in their music, don't you? Perhaps it's the influence of Asian music, one stroke on a Chinese gong and suddenly the winds are rushing in to your piece. I don't think you need to actually imitate the sounds of nature in your piece to get that." He sounds like a composer revealing his tricks or the hardware-store man explaining how to fix a drafty window. It's amazing that such a regular guy can write such remarkable music. Perhaps true sensitivity to sound is a quite ordinary quality that most of us have not tapped into.

"'La Mer' doesn't sound like an ocean, Debussy could just as easily have called it 'La Terre.' Nature is music but it becomes refracted in a curious way through the persona of the composer. I read once that Bartók had an incredibly acute ability to hear insect sounds. It shows up in his music, and it's so effective. Don't you think that happens, that we don't need to hear the actual sound? The *evocation* of nature is what matters."

Crumb didn't think up the idea of writing a whale piece himself. In 1970 a reel-to-reel copy of the humpback songs was already making the rounds among various musicians, and they were all amazed by it. The New York Camerata chamber ensemble commissioned Crumb to write a piece inspired by whale song, but Crumb didn't want to use the humpback recording in the piece itself. As a composer he doesn't believe in using prerecorded sound, because anything planned in advance takes away from the live musical experience.

"Each moment in great music is authentic, there's no accounting for it, something is there," muses Crumb. "If you use a recording you have no sense of bravura, there's no chance of falling flat on your face. It's what I call 'danger music,' fragile sounds that can collapse at any moment. The audience knows this is going on and they're relieved when a performer gets past a treacherous passage."

Crumb's music is famous for unusual techniques and theatrical moments, like a flutist playing a phrase over the piano's strings while the pianist holds the pedal down to create an instant sitar-like drone echo. He was the first to have classical players amplify their instruments so that very subtle gestures could be heard, like the tap on a clarinet key or the barest touch of a bow on a cello string.

So if he shunned the use of recordings, could he find a true use for the whale?

"The range was the first thing that impressed me," says Crumb, "from the pedal tones of the organ to sounds that go way beyond the limit of human hearing. A sense of musical phrase, an incredible composition that was going on, majestic, huge phrases. I loved the movement from the lowest to the highest sounds, the percussive elements, sounds like a thousand tubas playing at the same time."

Crumb didn't write a magisterial piece for a huge brass ensemble, that was more Hovhaness's line. Crumb instead used three of the four instruments of the Camerata: flute, cello, and piano, all amplified, demanding enhanced techniques from the performers. The flutist had to sing into her flute as she played, making some buzzing tones that mimicked the whale's sound. The cellist had to play high overtones beneath the bridge. The pianist had to "prepare" his instrument by sticking all kinds of screws and tape on the strings.

The whole conception of the piece bespeaks of evolution and of the sea, passing through a series of movements borrowed from the march of geologic time: Archeozoic (announced in the score as "timeless, inchoate"), Proterozoic, Paleozoic, Mesozoic, and Cenozoic (where humanity emerges, "dramatic, with a feeling of destiny"). And did I mention? All the performers have to wear black, half-masks over their faces. They are only removed for the final movement, the Sea Nocturne.

Sitting in the corner of the room the terrier paws at the door, yap-ping. "*Yoda, be quiet!*" snaps the master musician, and he returns to remember what his piece is like. "The sounds are epic, huge. Lions may roar but this is a much more vast conception with many more components than the howls of wolves or bears," says Crumb. He also made specific use of the idea that whale songs reverberate across the world.

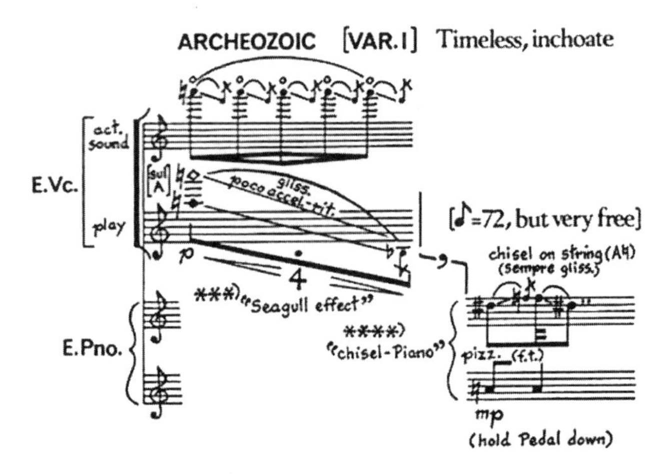


*Fig. 2. A fragment from George Crumb's score to "Vox Balaenae"*

"My piece sounds at times like it's under water, it's got something submarine. I mean little things, extra things like the seagull sound which I borrowed from Bert Turetsky. He invented that on the double bass and it works just as well on the cello, like fingering a false harmonic. It phases in and out, no one knows exactly how it works." Figure 2 shows a fragment of the score where the electrified cello is playing that very whale-like sound.

"Vox Balaenae" is clearly an homage to Olivier Messiaen's "Quartet for the End of Time," which may be the most important classical piece ever inspired by bird songs. Messiaen was certainly more wrapped up in the sound world of birds than Crumb ever intended to be with whales—he was never out on the sea transcribing the songs as they welled up from the deep, the way Messiaen did in the wilderness with birds. But Crumb also wrote a piece for a very small ensemble with a genuinely epic reach, mirroring the whole evolution of life on Earth. He thought about the grandness and extent of the whale's music, and he did not copy it or easily insert it into his own work. Instead he evoked it, not only with buzzy timbres and irregular forms, but by inhabiting the submerged magic.

Every effect and strange sound proceeds sequentially, so we can really hear the weirdness of it, much like the relenting solo explorations of a singing humpback whale, who introduces one astonishing phrase after another. We listeners drink in the sound. Crumb's succession of beauteous effects is serene and delicate, in that sense nothing like a 150-decibel whale. He sighs and takes another sip of tea. "Some people tell me my music is too *quiet* for today's world," he tells me. "It just gets lost in the noise."

It is a stirring, unusual, but ultimately accessible piece, with powerful tonalities luring you in to an alien world. Donal Henahan reviewed the second performance of the piece in the *New York Times* in 1973, impressed by the "fragile harmonics" and fuzzy piano preparations. He praised the delicate poetry of the sounds but cautioned that "sheer surface beauty of this kind is always suspect in intellectual circles. 'Vox Balaenae' may not turn out to be a major work. But don't bet on that." His second hunch was right on. It is Crumb's most-performed work, and it too has been performed for whales a few times. Their critiques have not so far been noted.

Crumb considered his music to come from a solid twentieth-century history of emotional energy. "Four composers were extremely influential on me, Béla Bartók, Debussy, Charles Ives, and Gustav Mahler, and they all were heavily affected by nature, they were pluralists in their outlook, so nature had a place. They weren't insulated from what life is and what the world is all about. Not like those middle-of-the-century purist abstract guys." Much of twentieth-century classical music lost this raw power. "It became university music, it had no potency, it became inert. It's music without sex, emasculated."

Since Crumb taught for years at the University of Pennsylvania, I wondered if anyone pressured him in that more standard, academic direction. (I remember they tried to pressure me, as late as the 1980s.) "No, even though much of the academic world was buried in gray music, we could ignore it. But today there's been a complete change, composers are hearing *sounds* again."

George Crumb showed us how the specific inspiration of one fragment of nature can completely change the way we listen. As computers make the manipulation of sound for its own sake ever more facile, people start to pay more attention to the immense musical power of raw, natural sounds. Today's music students study the perfectly realized splendor of "Vox Balaenae" as an example of how traditional instruments can be stretched to the brink of possibility. Rest in peace George, we lost you in February, 2022.

These underwater ballads appeared to human ears from nowhere, and musicians from all walks of life almost felt obligated to respond to them. Back in the early seventies, it is astonishing that so many leading music groups wanted to weigh in on the whale situation. I couldn't find a recording of Pete Seeger's reputed "World's Last Whale," but there are plenty of others. Country Joe (and the Fish) sang "Save the Whales!" David Crosby and Graham Nash, with James Taylor as backup, recorded an unusually complex arrangement in "To the Last Whale: Critical Mass" with wordless choral echoing, a tiny bit of actual whale song, and a plaintive lament for the great beast: "Maybe we'll go / Maybe we'll disappear / It's not that we don't know / It's just that we don't want to care." The faintest glimmer of whale song appears at the end, and we know we have to change the world. The band Yes did "Don't Kill the Whale." Even the Partridge Family devoted an entire episode to using their music to save the whales so we could save ourselves. Almost all of these songs offered the same basic message, but everyone felt the need to make it their own.

Captain Beefheart went a little farther. "I'm an animal," said this punk-like figure whose Magic Band sometimes veered in the direction of avant-garde jazz. "I think I sound more like a whale than I do John Coltrane." A more bluesy, raggedy version of Frank Zappa, Beefheart was suspicious of presenting whale songs themselves as music because he respected the animals too much.

"Whew . . . a thirteen-and-a-half-pound brain! I wouldn't record the whales though. I just want to help them, man. I mean, can you see a *whale* as a rock star? You know there is a family of whales, *heh heh*, in Trinidad Bay near where I live and when we play the music, the album, late at night when it's quiet they show up, they come and they dance!" When critics complained that his band sounded all out of sync like some garage free-for-all, Beefheart just smiled and said we just start and stop when we want to, like those whales. Their music "is past trigonometry, calculus, past polygraphs, and beyond that. They're smart, and it's frightening that we're killing them."

He handed out pamphlets for whale conservation groups like Project Jonah at his concerts during the seventies. He sensed something whale-tastic in the way he made music, raw craziness with a deep respect for the whole animal world. "Gonna grow fins," Beefheart wailed, "take up with a mermaid, and leave you land lubbin' women alone." No other pop figure is so suspicious of the human world—he's spent the last few decades painting alone in the desert, having left music behind long ago.

Over the years the songs inspired by whales appear less frequently but are perhaps deeper. One of the best is by Lou Reed, who turned "The Last Great American Whale" into a hero for our time: "He measured a half mile from tip to tail / Silver and black with powerful fins / They say he could split a mountain in two / That's how we got the Grand Canyon." Tougher than most songs of its kind, the lyrics admonish us for not having the strength to be concerned: "Well Americans don't care for much of anything / Land and water the least / And animal life is low on the totem pole / With human life not worth more than infected yeast. . . . They say things are done for the majority / Don't believe half of what you see and none of what you hear / It's like what my painter friend Donald said to me / Stick a fork in their ass and turn 'em over, they're done." Only the tiniest glimmer of a whale song appears at the end of the track. When Reed wrote about his early days in the Velvet Underground, he could have been talking about whales: "We heard our screams turn into songs, and back into screams again."

Laurie Anderson did a theater piece called "Songs and Stories from Moby Dick" in 1999, which became an album called *Life on a String* the following year. Again, you hear only a slight imitation of the whale, with evocative, searching texts. In one piece, "One White Whale," she wonders if the song will lead us to our goal: "How to find you, maybe by your singing / A weird trail of notes in the water / One white whale in all these oceans / Slipping through the nets of silence." We know so little yet pretend to know so much. The search is endless, and even when we find him in another tune, "Pieces and Parts," we still can't figure him out: "It's easier to sail around the world in a coffee cup / Than to see a whale when he comes rising up / We see him only in parts / A fountain, fins, a speck on the horizon / Giant teeth, an open mouth / Look out, look out, look out, look out."

Every musical portrayal of whales seems to combine longing and sadness. We want to reach them but we cannot. The sum total of human effects on the planet may have made the timing too late. Ecological longing continued to inspire more music. Other classical composers got on the cetacean bandwagon after a few years. John Cage, pioneer composer and musical philosopher, who for years urged us to enjoy sound for itself and to let art imitate nature by manner of operation, wrote his "Litany for the Whale" in 1980. This composition is a nearly half-hour piece for two male voices, each singing the letters of the word "whale," like "*woo ha el eh*," each to the other, never overlapping, like the actual solo humpbacks never interrupting each other across ocean miles. The whole work gives a meditative sense of wide spaces and great distances.

Toru Takemitsu, perhaps Cage's Japanese counterpart, met Scott McVay in 1970, but it took him eleven years to offer up a whale piece, specifically commissioned by Greenpeace in 1981. Takemitsu was also inspired by Melville's *Moby Dick*, impressed that, for once, a whale triumphed over its hunters in the end. Yet his favorite reference in the novel concerned the musical power of the sea itself: "Let the most absent-minded of men be plunged in his deepest reveries, and he will infallibly lead you to water. . . . Yes, as everyone knows, meditation and water are wedded forever."

"Toward the Sea: Moby Dick" is a swishing, wave-like tone poem echoing Debussy's "La Mer." Takemitsu adds some of the unusual flutter-tonguey flute sounds that Crumb so eloquently brought to whalify his own piece. Still, Takemitsu's work sounds more conventional, more derivative of these earlier masters. But the significance is that he, the acknowledged master of classical music in Japan, cooperated with Greenpeace to save the whales instead of continuing to offer humpback sushi on the menus of the nation.

You would think that jazz musicians would be among the first to welcome the strange culture of the moaning whale, yet it wasn't until 1979 that a jazz player explicitly emulated whale sound in his own playing. With just the right amount of undersea reverb, bassist Charlie Haden begins "Song for the Whales" with an improvised melee of descending bowed notes. He is known for the deep sensitivity of his tone and attack. His band Old and New Dreams was composed of four of Ornette Coleman's most distinguished sidemen. They continued Ornette's style of playing free jazz with an undeniable sense of groove and swing.

Don Cherry interjects with dolphin-like pocket trumpet *upbleeps*, and eventually the two remaining members of the band, drummer Ed

Blackwell and tenor player Dewey Redman, come in with a sped-up dirge with a fast running drum line beneath. Whales come back up on land once more, and this time they dance. The structure of the piece is unlike anything else the band ever did, once again showing that when musicians take up with humpback song, their music changes and there's no going back.

The one jazz musician who has most allied his work with the sounds of nature is Paul Winter, a man who has used wolves, whales, and birds throughout his work. Winter was discovered as an up-and-coming young saxophonist while still in college in the early 1960s. The U.S. State Department sent his band on several international tours where he got acquainted with Brazilian music, which he soon integrated into his sound. In 1962 Jackie Kennedy invited the Paul Winter Sextet to be the first jazz group to officially perform at the White House.

In 1968 a friend of Paul's said he ought to put on headphones and check out these whales, saying it was way better than an acid trip. "Their voice was sort of a cross between an elephant trumpeting and Miles Davis. They had this bluesy quality that was so poignant. This made me realize that there is perhaps a universal yearning that is shared by all species, this calling, crying quality, in their singing." Winter may have been the first to think of whales as subterranean jazz musicians.

He was inspired to attend a lecture by Roger Payne, and from that moment he was hooked. Few speakers on whales are as gripping as Payne, who has brought thousands to tears with the songs and stories of the humpback whale. Through his research and activism he has done more to spread the beauty and dignity of whales and their sounds than anyone else.

Winter was astounded by Payne's diagrams of the deep structure of the humpback song that resounds for up to thirty minutes before it repeats. He immediately felt an extraordinary intelligence behind the songs of these animals that were being killed for lipstick and dog food. Paul asked Roger what he, as a musician, could do to help. Payne said, "Make sure nature has a place in your music."

By the mid-sixties Winter had moved to the countryside of Connecticut, not far from where Dave Brubeck, Gerry Mulligan, and Winter's first producer John Hammond lived. Taking to the woods is a challenge for a rising jazz musician. Jazz is famous as an urban music—the jarring juxtaposition of city sounds suggesting new sudden moves, improvisations out of the melee of people meeting, cultures blending. Living amid so much water, wind, and trees, Winter began to wonder how a sense of place could meld with his sound.

The first nature piece he did was based on the beautiful Eliot Porter photo book *In Wildness Is the Preservation of the World*. The vivid, pure photography juxtaposed with pithy quotes from Thoreau set the tone for a whole line of Sierra Club Books that showed how beauty, not only worry and fear, could coax the public into the environmental movement. Winter's piece of the same title used a litany of voices of endangered species, with their names recited one after another: "Black-footed ferret, whooping crane, Alabama cavefish. . . ."

After one such performance, someone in the crowd came up and gave Winter some advice: "You're never going to accomplish anything by making people feel guilty. You need to celebrate the creatures instead." He had to learn to create the musical equivalent of a Sierra Club book. It took Winter nearly a decade to figure out how to do that.

In the early seventies Winter found himself influenced by another Connecticut musician, the iconoclastic composer Charles Ives, who had lived just up the road in Redding, fifty years earlier. An early champion of a raw, primal kind of Whitmanesque Americanism in classical music, Ives set up different marching bands clambering up and down hillsides, playing cacophonously together in and out of time. One might call him the Captain Beefheart of his day.

Winter tried in his own way to make a music rooted in his home landscape. He began to organize music villages where all kinds of people would get together and play, breathing in the clean country air. Following Ives, he saw a possibility of blending classical, folk, and jazz music. He worked with Peter, Paul and Mary as well as Pete Seeger. He gathered together some of the best and least conventional improvising musicians, guys like guitarist Ralph Towner and oboist Paul McCandless, who applied a personal, jazz-type sound to classical instruments. Winter called his band the Consort, evoking a purer world of medieval troubadours.

Their album *Icarus* was produced by legendary Beatles producer George Martin in 1972. Mixing Bach, drums, ancient ballads, and wild soloing, it presents a genuinely optimistic, new sound. Martin has famously called it the finest record he ever made. The title song was even launched into outer space on one of those Pioneer missions that left our solar system. (Also included, at the behest of Carl Sagan, were songs of the humpback whales themselves, just in case the probe is found by an alien intelligence that might understand them better than we do.)

*Icarus* is a beautiful record, and it was critically well received, but it did not become a hit and did not bend the jazz world in Winter's nature-oriented direction. He went back to the drawing board. "The failure of that album to connect with the culture led me to withdraw and want to write my own music."

The Consort started to play ecology movement benefit concerts, and Winter began to meet more activists and biologists. He went out with David Mech to listen to wolves, and discovered that his especially soft and pure saxophone tone seemed to blend perfectly with their howls in the night. He first saw gray whales from the Greenpeace boat, the *Phyllis Cormack*, off Vancouver Island. "I had never before seen a whale, I had only the sounds and this vague notion of what they are. What struck me so deeply was this slow-motion grace, their surfacing, this powerful spout, and then they would dive. Suddenly I understood a whole different aspect, not just this thunderous power."

The next day he took a cheap saxophone out on a Zodiac raft and tried to play for the big, mostly silent grays. Although there are famous pictures of Winter playing his sax out on the water with whales, he's never been sure they could hear or respond.

In the seventies, Winter went down to Baja California three years in a row, where the gray whales in their winter calving lagoons have since gotten more and more interested in human whale watchers. Someone was trying to make a film of people and whales getting closer together. "There was a yoga dancer on the beach, and I was playing my soprano, and both of us were completely naked, and they had a big plastic inflatable whale on the beach and some real whales in the water watching us. This film was never completed; I often wonder what happened to the footage. And I wonder what the whales thought of it all."

(Now it so happens that a few months after visiting Paul I met this very same dancer on a beach on Maui, about thirty years after the fact.

"Naked?" she laughed. "I would never do those dances naked. They're sacred. Of course, there was a lot of general nakedness going on down in Baja, Paul's memory may be confused.")

Probably because of tales like that, and because of the softness of his speech and his saxophone sound, Winter is often thought of as a father of New Age music, a category in which he has won many Grammy Awards. To those who support it, the New Age genre refers to that kind of sound that connects the listener to higher spiritual states, into a kind of pan-religious bliss of a hopeful future. To those who don't like it, it's trumped-up muzak or milquetoast instrumental folk. Winter, though, has always been a jazz musician, one who has tried to push the boundaries of that genre to encompass the tunes and inspiration of many species through new clues for improvisation.

Since his performance in the White House fifty-odd years ago, Winter has continued to be a musical activist, creating beautiful and provocative soundtracks to numerous important environmental and political campaigns. His 1976 album *Common Ground* is the first to incorporate the melodies of the creatures themselves, specifically the trio of whale, wolf, and eagle, where the specific tones of each creature's tunes form the basis of the music in a direct and easily accessible way. Paul McCandless improvises with an eagle's cry in the loosest piece on the record. The wolf howls over a minor, tragic harmony in "Wolf Eyes," and the whale piece is called "Ocean Dream."

The piece opens with the hissy waves of the original recording from Payne's *Songs of the Humpback Whale*. Organ and guitar enter, with a downswept sigh. Winter sings a melody that heads up a tritone, the inverse of the whale's own interval: "Ocean child, come now home, holy wonder, holy one." The whale song comes back, weaving in and out of the human melody as the cello picks up the anthem as more squeaky whale songs emerge. The whale recedes into the distance, the saxophone drifts back. A wash of descending weird washes. They sound electronic, but they're actually humpback wails.

Hearing that record at age sixteen changed my life. When I first heard Paul Winter, I was amazed that music and ecology could be honestly combined. I imagined that better listening might really lead us all closer

*Fig. 3. Humans and whales in harmony. From* Mind in the Waters.

to nature. I even named my high school band Ecology. Winter showed a way that musicians could contribute to environmentalism not just through propaganda, but by teaching people how much music could be found in nature itself.

"That was the first time I ever tried to sing," says Winter. "And almost the last. It was fun, but I never felt like it was my essence. Lyrics to me are limiting. You can't take people into the realm of magic and mystery when words are in the way. That's just the prejudice of a life-long instrumentalist. I'd much rather aim for something that's universal, not limited to one language." Winter's singing voice is soft and alluring, a fine complement to his horn playing, a classical sound that few other jazz players use as effectively. Certainly influenced by the Paul Desmond tone that was so much a part of Dave Brubeck's sixties quartet, it is not a sound that has spawned many imitators today. Perhaps, like George Crumb's music, it is too quiet for the noise of the modern world.

I ask Winter what he thought of the works of Hovhaness and Crumb, both written at the dawn of human interest in whale music. "That's head music. I have always been interested in *heart* music." Makes me smile, because that's just what Crumb said about his nemeses, the chilly composers of atonal serial music like Schoenberg and Berg, so beloved by the academy. Perhaps for musicians, whatever we embrace comes from the indescribable heart, while whatever leaves us cold is the work of a mind more willing to calculate than to feel. But Winter is more practical in his rage. He wonders more about the fact that most musicians didn't give whale song a moment's thought:

"What about all the people who *didn't* get inspired by the whales? There should have been *thousands* of compositions." This should not be considered a mere historical fad, a relic of the 1970s. "Our only hope at salvation is to recollect with the wild world around us and to rejoin the family of life. Not just for aesthetic reasons or to preserve them, but because they are our elders, and we need to learn from their wisdom and their example of how to live in the world without defiling our own home."

If Winter sounds a little pious here, you may find it no surprise that for nearly three decades he was "artist in resonance" at the biggest cathedral in the world, St. John the Divine in New York City, where his summer and winter solstice concerts draw thousands of listeners over many nights. There is nothing quite like hearing the jazz of nature echoing in this giant church. Those who call Winter a New Age musician miss the point, that he is trying to reach beyond humanity to the hope of reconnecting our marauding kind with the gentle voices of the eternal, natural world. He is actually seeking something sacred.

Not thinking of himself as a singer, Winter rarely performs "Ocean Dream" today. His next record, *Callings*, in 1980, was the first on his newly founded Living Music label. Winter wanted total control of the recording process on this, a concept album featuring a sea lion pup named "Silkie" on a fantastic journey from Baja California all the way to Magdalen Island off of Prince Edward Island, where harp seals are slaughtered for their fur. This time Winter went back to Payne's original recordings and heard something that sounded like a lullaby: "I had this fantasy, that this song, from one of the largest mammals in the sea, was crying out for the fate of these small, helpless creatures. . . . It was an act of reverence to put these humble human harmonies beneath this whale melody."

This soft, gentle song, with the rather weighty title "Lullaby from the Great Mother Whale for the Baby Seal Pups," is a moving melody even if you might find the tale behind it a bit heavy. This is the most frequently performed piece of Winter's whale music: a simple ocean melody turned into a chorale, with clear, honest harmonies.

In fact, it is based on the same fragment of whale song that Judy Collins used in "Farewell to Tarwathie." The humpback fragment comes from *Songs of the Humpback Whale*, and this particular song was recorded in Bermuda in the spring of 1967. Collins used it in 1970, Winter in 1979. The whales change their song as a population from year to year, so forty years later none of them are singing the same songs anymore. None except for the Paul Winter Consort, that is. You don't hear much whale song in classical, jazz, or pop music today. Perhaps we're all so used to it that it's become a cliché. Did these many efforts help at all to save the whales?

For Katy Payne, who did so much of the most sensitive work really listening to and trying to decipher the song, the most important thing was how this whale music made us reconsider all of nature: "There was a burst of realization that the world could change its relation to wildlife. The reaction people had to hearing these sounds made whaling obsolete!" Whaling had been the only reason people knew anything about whales. As people heard the songs the desire to kill whales soon lifted away.

Not that it was so easy to convince the governments of the world that this was a good idea. It was public concern that forced our leaders to suggest that whaling must be stopped. In June 1971, the U.S. Senate unanimously passed a resolution demanding a ten-year halt to commercial whaling. At the time the government's highest-ranking whale scientist, director of the Smithsonian's Marine Mammal Council, was Carleton Ray of Johns Hopkins. This is what he had to say: "I don't find it very relevant to hear that whales produce music. Cock-a-doodle-do produces music too. Whales are smarter than chickens, but it is not relevant to say that whales have a complex social life. So do all animals, including the cows that we eat." Ray was also famous for arguing that clubbing seals to death was a quick and humane way to kill them.

Despite his academic pedigree, public opinion turned against Ray. Whale songs were all over the airwaves in the early seventies, and Congress succeeded in passing the Marine Mammal Protection Act of 1972, placing the United States at the forefront of global environmental protection. It is still one of the world's most exemplary articles of law concerning whales, sending a clear message that cetaceans are to be studied and respected, not used. Section 101(a): "There shall be a moratorium on the taking and importation of marine mammals and marine mammal products . . . no permit may be issued for the *taking* of any marine mammal and no marine mammal product may be imported into the United States."

Whales got a great boon of protection with this law, and the fifteen years of public outcry and music inspired by the plight of the whale helped to push global sentiment in their favor. Greenpeace became famous for its Gandhian techniques of putting their rafts right between harpoon and whale. The elegant book *Mind in the Waters* was published by the Sierra Club, an anthology of whale tales, drawings, and scientific reports, which drew further support for these wonderful, little-known animals.

By 1986 the International Whaling Commission had to respond to the overwhelming support of the world's people, and their governments, for a moratorium on commercial whaling. Though Japanese, Norwegians, and Russians continue to hunt whales and defy the ban, world public sentiment is against them and protests continue. Whale populations have been recovering steadily since the ban, and these countries sometimes argue there are more than enough whales out there to kill—no one will notice and whale watching can still thrive. But most of the world's people now consider whaling to be an obsolete and barbaric practice. If they want to openly continue to kill whales, it's up to those countries to prove otherwise.

"There was one moment in the seventies in which I most admired Roger," smiles Katy Payne about her ex-husband. "At a meeting in Bergen, Norway, with hundreds of people there deeply invested in the whaling industry, he simply told them that whale watching was starting to make more money than whale killing. When people began to realize that this is the case, I realized, the world is changing. And it was the song that did it.

"But people forget quite soon; there's this love for novelty which probably drives the changing whale songs, which also drives people's interest. We find fashion a really important part of human culture. Our culture got really fond of humpback whale songs, and then they forgot about them."

Until now. It's time to bring whale song back into the human world again. The long, epic rhymes of humpbacks. The *tap-tap*-tapping of sperm whale click trains. The cacophonous free jazz of belugas and the kinship whistles of orcas. The thousand mile thrums and beats of blue and fin whales, crossing whole oceans in less than an hour.

What about that Pete Seeger song? In 2007 there was no recording of it, but a quick internet search did conjure up the lyrics:

It was down off Bermuda
Early last spring,
Near an underwater mountain
Where the humpbacks sing,
I lowered a microphone
A quarter mile down,
Switched on the recorder
And let the tape spin around.

I didn't just hear grunting,
I didn't just hear squeaks,
I didn't just hear bellows,
I didn't just hear shrieks.
It was the musical singing
And the passionate wail
That came from the heart
Of the world's last whale.

This song seemed quite different from the others of its time. It's all about how the song is recorded, and how the music happens. At the end, it does return to a morality tale:

So here's a little test
To see how you feel,
Here's a little test
For this Age of the Automobile.

If we can save
Our singers in the sea,
Perhaps there's a chance
To save you and me.

I heard the song
Of the world's last whale
As I rocked in the moonlight
And reefed the sail,
It'll happen to you
Also without fail,
If it happens to me
Sang the world's last whale.

Seeger got all the details right: the microphone deep under the sea, the rocking, rhythmical beat of the boat swaying back and forth, and the whale poetry resounding and repeating underneath. Never recorded? I was shocked. Pete Seeger lived just up the road from me, so I wondered if I might rectify that situation. Let's record it today.

We had recently performed on the same bill in Toronto, so I gave him a call. "You remember that song about the world's last whale?"

"What song?" The scratchy voice on the other end of the line sounded suspicious.

"Goes like this: 'I heard the song, of the world's last whale. . . .'"

"Ah yes, you know my mind doesn't remember it, but I believe in muscle memory. My body's still got that tune."

"You want to sing it?"

"I'm eighty-seven years old—too old to sing. But *you*, you should come on down to the Hudson riverfront and play some of those whale songs of yours while the swimmers cross the river from the other side. They're going to love it."

"I'll do that if you sing the song."

"What song?"

"'The World's Last Whale.'"

"Oh, we'll see about that."

The following weekend I ambled down to the waterfront festival in the nearby town of Beacon. Pete started the Great Newburgh-to-Beacon

Hudson River Swim four years ago to remind us that this river has gotten clean enough to jump in. He wanted to celebrate some environmental good news and raise money to build a lined swimming pool in the river to make it safe enough for all.

I had done my homework, and found out that in 1988 a humpback whale had actually swum up the Hudson River. Why not play some of its songs to inspire swimmers just finishing their mile-long crossing? I asked a question of the crowd, "How many of you know what a whale sounds like?" The parents and grandparents smiled.

Pressing "play" on my computer, the swoops and bowed bass notes resounded from the speakers on stage. I accompanied on clarinet, trying to play sounds that would blend. It was a bit of a change from the usual river festival folk tunes, but the swimmers and their families didn't seem to mind.

A tall, rail-thin man with a beard pushed his way to the stage with a banjo and a big pile of papers. "You know, I had totally forgotten about this song until this young man brought it back to my attention," Pete nodded in my direction. "Here are some copies of the words, and I wrote out the music, too. These whales still need our protection. Anyone who wants to keep this song alive, here, take a copy."

Could Pete Seeger still sing after sixty years on the road? More than once I've heard him go on unaccompanied for an hour at a time. On the tribute album to Seeger's work, I find that Bruce Springsteen, with his worn, gravelly delivery, sounds a lot older than Pete. On one of his final records, *At 89*, Seeger did end up singing this song of his that I dug up on the internet. I believe it even won a Grammy. And. Now you can find plenty of versions on "The World's Last Whale" on YouTube, from all over the world. Rest in Peace our greatest folk singer and musical activist; music will not see the likes of him again for a long, long time.

We're not going to lose that song now. This lilting, grooving tune in a doleful key reveals exactly what the song of the humpback whale meant for us when it first became known: in the 1960s, miraculous underwater recordings revealed there is music under the sea, and we learned of one more rare thing of nature that was fading away. If we don't work hard to save this song that is so radiant yet also fragile, we're going to disappear just like the whales. It's a simple moral from a beautiful sound.

# 3

# THOSE ORCAS LOVE A GROOVE

## Making Music with Killer Whales

Can humans make live music together with whales? Modern technology brought us the whale song, and it can send our music down into their depths. You already know that hydrophones can record the strange behavior of underwater sound, but here's the rest of how it's done: sit on the deck of a boat, or on shore if the water adjacent is deep enough to welcome whales. Play your instrument, say a clarinet, into a small microphone on a stand. Plug that microphone into a preamp. Send the output to a small battery-powered amplifier. Send the output of that amplifier to an underwater speaker, something that used to be expensive but is now cheaply available from China. Send that clarinet sound deep under water, where even with low power it can travel quite far. Record the output from the hydrophone, and you will hear your underwater self, along with any whales that happen to be singing.

It's not going to be as easy as jamming with birds, because usually I will not be able to see who I'm singing with. I'll play by ear, and whatever music I make with the whales will not sound entirely human. Some whale sounds are so low or so high that we cannot hear them. Reaching beyond species lines always means extending the boundaries of what counts as music. You will hear things that surprise you, and play sounds that you would never play otherwise. A strange beauty emerges even if the message cannot be decoded.

After dolphins, the whales we know best are orcas, or killer whales, so named because their diverse populations will attack and eat almost anything that lives in the sea. But they also have a gentle side. They're curious about people, and for decades whale watchers, scientists, and musicians have felt a special connection with these animals. Some have been watched so closely in the waters surrounding Vancouver Island that each individual whale is known by name, family, and lineage.

The whales can be identified by their unique tail and fin patterns, and since they are black and white, they are much easier to see than any other whale species. Orcas are social animals and use sound to keep the pod together as they zoom through the sea. They use song in a way that more closely resembles the laughing thrush than the humpback whale: the guidebooks published to help visitors identify each whale emphasize their appearance, but the original difference between pods and clans of orcas was discovered through sound—first by listening, and later by interacting with them.

Who has played music to killer whales? Paul Winter, for all the great photos of him and his soprano sax standing in rafts playing pure tones out to leaping orcas, says he only tried it once, thirty years ago, blowing the sax into a big metal tube that stuck into the water. He heard the whales, the whales heard him. "At one point it seemed like they were responding," he says. Winter included only a small taste of this duet on his *Callings* album.

A few scientists in the sixties and seventies played music to orcas, but they don't much like to talk about it today—seems a bit New Agey by today's standards of rigor, or at least beyond the limits of experimental control. Paul Spong, the guardian of North Vancouver Island's orcas, and veteran of several decades of research into their habits and behavior, began his work on the sensory systems of orcas at the Vancouver Aquarium in the mid-1960s. Getting food in exchange for tricks did not interest the captive whales, so Spong tried playing a whale back a tape of his own sounds. Not interested. He tried the sound of another whale. A faint glimmer of curiosity, but that didn't last. The aquarium whales seemed bored. Spong wanted to make the whales happier in their confinement, and he got interested in what animal trainers call *enrichment*, making the lives of their charges just a bit more interesting.

When a whale named Haida lost his mate, he sat despondent in the pool as if he too wanted to die, refusing to eat anything for a month. Spong invited jazz flutist Paul Horn, known for his solo recordings inside the Great Pyramid and in the Taj Mahal, to offer some gentle music to cheer up the bereaved whale. For a week he played mournful elegies for the whale's lost love, but Haida showed no interest. One of the trainers urged Horn to try a more positive approach.

"I get the feeling you are playing rather sad music," she said, "reinforcing Haida's unhappiness. Maybe you should try providing some positive energy to bring him out of his depression."

Horn leaned over the water and talked to Haida eye to eye: "Look, I've been coming here for three days now and you have totally ignored me. I get no response from you, and I'm getting *bugged*."

The whale said nothing. Horn continued. "We know you've suffered a great loss, and we sympathize with you. But thousands of people come to see you, they respect and admire you, but you're letting yourself down.

Not only that, you are letting yourself down and you're letting life down. Life is a very precious thing, Haida, so get your act together and snap out of it. I'll come back one more time. If you don't respond tomorrow, then I won't come back again."

The next day Paul Horn was back at the pool, playing the same mournful music he always played. But right away Haida moved his head, the first movement he'd made in weeks. Horn walked around the pool, playing his flute with conviction, just like he'd famously done walking around the Taj Mahal. For the rest of that week the whale followed the flutist, and after five days, Paul put his flute down and dangled a herring over the water. Haida hesitated a moment, but then decided to scarf it down. It was the first food he had taken in a month, and the first time a whale was known to have been brought out of depression through music therapy.

Haida recovered enough to be brought a new mate from another oceanarium. The next time Paul came to play, the female whale seemed more interested in the music than Haida, who disappeared to the other end of the pool. After a few minutes hiding deep down, he suddenly leaped out of the water at high speed and smacked down on the water as hard as he could, spraying a huge wave all over Horn and his precious flute. "It seemed clear he was either jealous of me or annoyed that his girlfriend was more interested in the flute than she was in him. Whatever the reason, he was pissed off and got me soaking wet. This was the last time I saw Haida." Music later became part of the general enrichment program for captive whales all over the world.

But Paul Spong was not satisfied. He had grown confident enough to touch the huge beasts. A whale named Skana seemed to enjoy being touched and rubbed, especially by Spong's bare feet. One morning her behavior changed. Spong was sitting with his feet in the water of the pool. Skana approached very cautiously, then opened her mouth wide and dragged her teeth quickly across the top of his feet. He pulled his toes out as quickly as he could. After a moment, he put his feet back in. The whale came back, again flashing her teeth. The feet came out again in an instant. They did this many times in a row, until finally Spong was able to control the urge to flinch. When he stopped reacting, Skana stopped with the teeth.

Spong's attitude to these animals suddenly changed. It seemed to him they were conducting their own experiments on him. How could he condone keeping animals with such intelligence in captivity? Much of what we have learned about whale bodies and brains could only be found out with captive animals, but Spong could do this work no longer. The more he learned of these animals, the more he wanted to offer them freedom. He made preparations to leave the aquarium. He believed the whales should be in the wild, and so should he.

In 1970 Spong headed to Johnstone Strait to set up a field center for the study of wild killer whales. It is still going strong on Hansen Island at the end of the strait, the longest continuously running whale field research center in the world. Since these animals live so close to human habitation in clear social groups, they are much easier to study than the humpbacks who migrate far across wide oceans, who have the longer, more epic songs. But orca songs are equally interesting because they are so *interactive*—these whales use a distinct sonic repertoire to stay in constant communication with each other. So it's more likely that a killer whale will respond to a new human sound than the staid, solo-performing humpbacks.

In the beginning, Spong played all kinds of recorded human music through underwater speakers to see if the wild, free-ranging orcas showed any interest like their cousins in captivity. They didn't seem to care. Then he tried live music, with a Vancouver rock band called Fireweed jamming from a fifty-foot sailboat. One pod assembled around the boat and followed along for several miles, something that hardly ever happened. The younger whales especially seemed to have an ear for rock and roll, and they stayed close to the boat as long as the music was blaring.

The next summer Spong wanted to get closer. He took a solo kayak out to be in the midst of the pod, and, perhaps inspired by Horn, brought along a wooden flute. Playing soft tones above the water, he sensed the whales could hear the sound under the surface, even though he couldn't hear them. After building a sense of trust with the orcas, he began to swim and dive among them. In *Mind in the Waters*, Spong writes his most moving words on why music must matter to killer whales: "Sometimes, particularly on a still night, a pod or part of a pod, or perhaps just a single whale, will hover offshore for an hour or more, apparently tuning in to the music. Sometimes they seem to join in the celebration with the chorus of their voices and the dance of their bodies, visible to us from the bubbling phosphorescent wakes they leave behind."

Others were drawn to the area by such testimony. Erich Hoyt came to Johnstone Strait in 1973 to make one of the first documentary films about orcas. He got ahold of some of Spong's tapes of orca whistles, and started learning the repertoire on his Arp Odyssey synthesizer, an early electronic analog instrument the size of a portmanteau suitcase with a multi-octave keyboard and about forty sliders. It was particularly adept at making loud, strange whoops and pure sine-tone whistles.

Hoyt moved the faders up and down, twiddling knobs, adjusting timbres. When he first got out in the field to hear the actual whales, he felt he was in the hall of nature listening to a grand new piece: "I've just walked into the opera house, I have no program. Strange new players are premiering a piece by a flamboyant new composer." Swelling, discordant strings morphing into rusty saxophones. Pizzicato trumpets? Impossible echoes? The reverberations carried for miles.

Hoyt sat at the Odyssey's controls, waiting for his cue. Suddenly he heard a familiar phrase, one he thought came from a whale named Wavy he had heard a few days before. Other whales answered the same phrase, always with slight variation. The accent switched from the first to the third note. It was an ensemble piece, with hardly any space for him. Finally the moment came, he played the phrase he had prepared, but he changed the beat just a bit. A few seconds, then he heard it: "A perfect imitation of what I had just played to them—in harmony! They did not repeat their own sound; rather, they duplicated my human accent." The first time was the best time. When Hoyt repeated his experiment, the whales had lost interest. Perhaps they don't care that much for our music after all.

Almost as quickly as it sprang up, the idea of playing music to orcas began to fade away. As Spong became more of a scientist, and Hoyt became more of an activist, they were cautioned by their peers against doing things so risqué as jamming along with whales. It didn't sound like serious work. It wasn't research, because it was never done in a controlled manner. Animal behavior science is hard enough to control as it is, and with music in the mix, could a scientist hope to be taken seriously? The standards of whale science were getting more rigorous, especially as the formerly flourishing orcas became increasingly threatened, even in the apparently pristine environment of Johnstone Strait. Shipping traffic was increasing, pollution was on the rise; loggers would happily clear-cut every one of these verdant islands. It was time to focus on conservation.

Both these guys worked hard to set up marine sanctuaries for these animals, Spong putting roots down on Hansen Island, Hoyt working all over the world to save the whales. The two men have remained as firmly committed to the cause as anyone. But in the beginning they were both motivated by simple wonder. As Spong puts it, "The whales sang and called to us and we returned their voices with everything we could manage—flute sounds, imitation orca sounds, singing and laughing—in the joy that only free creatures and free people can create together." After a few years, neither of these two champions of the whale talked any more about the buoyant mood of those early days. With battles to fight, who has time for fun?

There is only one person who has remained dedicated to playing music live with whales long after it ceased to be fashionable. If anyone was going to show me how to make music with cetaceans, it would be a man I first saw on television in 1980. Just before Thanksgiving, I watched a thin, bearded figure playing flute in a pen filled with turkeys on a show called *That's Incredible!*

The flutist was playing an ascending scale very slowly, and on certain notes all the turkeys started to gobble in unison. Either they were responding to the music or they were nervous that they would soon be roasted as part of a human feast. For years this enigmatic image haunted me when I thought about music and nature, flutes and birds.

Only when I met him did I realize this was none other than Jim Nollman, who has pledged to play music across species lines for longer than anyone, inhabiting a porous area he likes to call "the charged border," where whales and humans meet. He describes it as a place you're going to live in and truly explore: "The charged border emerges as a luminous crack between worldviews—a place where mystics, biologists, retailers, historians, shamans, tourists, fishermen, and children gaze upon the same animal and yet take away entirely different meanings from the experience."

After years of following his work, I was looking forward to taking a trip with Jim to try and make music with the orcas of the Johnstone Strait. But Jim wasn't so enthused. Living in the San Juan Islands in a house and garden he and his wife have slowly built themselves, Nollman combines a Fritz the Cat swagger of having seen it all and done it all, mixed with a touching humility of believing his life has not gone where it should, that the world still doesn't get it, that the border cannot be breached.

"I don't know, man, I think I'm done with whales. The best singers died long ago. Don't think you're going to get much reaction these days. Too many whale watchers and scientists harassing those pods. Don't expect anything." Nollman has been filmed by outfits from all over the globe, and he has sold thousands of copies of his record of whale-human interactions called *Orca's Greatest Hits*, simply by force of will, traveling all over the planet, lending his presence to show that whales should be played with, not murdered for oil and meat.

Just the fact of Nollman showing up can bring the idea of intelligent, musical whales into the media. He once temporarily stopped dolphin slaughters in Japan, though such traditions keep stubbornly returning. In California he once worked with the artist collective Ant Farm, and was lauded by Governor Jerry Brown. John Lilly once sent him to Mexico for eight months to see if Nollman could break through to the dolphins' musical world.

Not that I am recommending that you try any of this on your ocean vacation! Playing music with whales in United States water is forbidden by the Marine Mammal Protection Act. No whale can be "taken," and that means no hunting, capturing, killing, or harassing of any marine mammal. The government considers music a form of harassment. Nollman's none too pleased that scientists, even some of the same people who used to take whale music more seriously, are now the only ones allowed to play around with whales in the name of gathering information. He doesn't like their methods and he doesn't like their arrogance.

"We wanted other people besides scientists and wardens to have access to these beings. In one place I went in the Arctic, a high profile, government-sponsored research project was monitoring beluga movements in an attempt to best determine which native people should be allowed to kill them. Strip away the hifalutin language, and you could see what those biologists were actually doing was driving individual whales up onto a beach with motorboats, and then using electric drills to punch holes in their backs to affix radio collars."

Nollman prefers drama to accuracy, a good story over the fastidiousness of the scientific way. "We anchored a boat in the same bay off of Vancouver Island for six consecutive summers, played music into the water every night at ten p.m., and recorded the musical interactions with the orcas. A well-known biologist was very interested in this particular project but lamented the fact that we never bothered to replicate our performances. I told him jazz musicians never play the same note twice." I wanted to visit that very same bay. I knew it would be an interesting ride.

We take the ferry from the San Juan Islands across into Canada, where the customs agent is a bit suspicious of the strange devices in our trunk. Eventually he lets us go, and we begin the long drive north up Vancouver Island. Now Jim is starting to worry me. "I'm not really interested in whales per se," he repeats as a raven squarks overhead. "It's all a vehicle. To get closer to nature. To be in nature—to get outside of myself." Late in the day we arrive at the end of the road, the tiny outpost of Port Hardy, the final town on the island, ensconced in fog. The next morning the sky is clear and we head out on a glass-like sea toward a distant commotion in the water. Sticking the hydrophone down we hear nothing but a faraway hum. "Must be a submarine." Jim tends to pick the most outlandish explanation for things involving whales and the sea. Soon we saw it was merely a distant cruise ship.

Sometimes it seems that Nollman will say *anything* if it might help save the whales. His writing is full of provocations: "The Kwakiutl and Haida believe orcas comprise an advanced culture a step above human beings. Is this step like the difference between a C sharp and a D? Environmentalists point out this is a species that does not attack human beings *even though it can*. Mystics explain it as the difference between humans and angels." Jim believes that whales are telepathic beings that humans can learn to tap into. He tells a story about four sperm whales suspended in the water with their heads facing each other like a cross, combining their power to send thousand-mile undersea radar out in all directions—or at least that's what a former CIA agent he met in Alaska once told him.

"Do you have a good recording of those big summer interspecies jam sessions?" I wondered.

"I was never very good at documentation," laments Jim. "The most important thing to me about this work has always been just doing it. We had the core group for years who would be ready to come out and play with the orcas at beck and call, and then the supporters and their friends who would give us the funds to do it, and they'd come along for the ride." He points south down the strait. "We had this place we called Orcananda, and we'd usually charter a boat from Seattle or Friday Harbor and head up there for weeks at a time. Around 1991 this work leveled off, it started to fade out. I began to realize we weren't playing with the whales, but playing with *two* whales. A mother orca named Nicola and her son A6." (Every one of these Vancouver Island whales has a letter and number, identifying which clan she is in and what her lineage is. This is the result of years of meticulous work by Paul Spong, along with Michael Bigg and John Ford.)

Jim continued: "Nicola was legend among the whale watchers up here. She would actively approach humans, really seem to enjoy singing along. Everyone had stories about her. In 1991 she died, of natural causes. After that her son lost interest in us. All the resident orcas still have names and numbers today, mostly codes, nothing personal. They may have been whalewatched just too much. And why should they care about us anyway? Since then the interactions have never been the same."

Nollman seems always to be longing for bygone days. "In the early nineties I stopped doing this work because I was not sure there was anywhere else I could take it. I started to write more about it. Sure, when I was your age, David, that's when I thought I had the world in my pocket. I was traveling around, making money, writing books. I thought I would never run out of things to say. Now, once more, I'd rather just play."

So play we will, as we take the Hardy Bay charter boat farther out to that glassy sea beneath tiny green islands. The map calls this place "God's Pocket." Here the fog rolls in again. We could be anywhere. Who's under water? Let's roll a hydrophone down and see what's up. There they are, the whoops and squeals. Orcas below us, the first ones I have ever heard. "Quick, get out the equipment," Jim commands. I hastily pull out my AA-battery-powered system of microphones, amplifiers, and the underwater speaker. Nollman shakes his head. "That thing is merely a toy meant to play disco in swimming pools. No whale is going to get off on that."

Jim heaves up his behemoth of a system, an impressive custom job designed in the eighties. The all-in-one mixer and amplifier weighs at least thirty pounds and runs off a boat battery. Jim calls it the Interspecies Sound System. I'll just call it the Beast. "Is red negative or positive?" he asks me. In the flurry of excitement I can't remember either. He tries one way, nothing happens. Switches it, and acrid smoke steams out of the machine, wafting up through the quarter-inch female jacks. "Shit, this thing is fried," he moans. Meanwhile the boat is drifting toward shallow water and the depth-sounder beeps. "I can't do this and keep track of the boat at the same time!" The Beast is dead, but the ship sails on.

My setup, as cheap as it is, seems stable enough to produce a pretty loud clarinet ring under water. I start with really high squeaky notes like what I'm hearing from the whales. We see nothing. I let out a high shrieky D. Then a G a fourth above. I hear those same notes back, then mixed with a whirr, and a buzz.

Jim winces. "Why are you playing those shrieky sounds, Dave? You're hurting my ears."

"That's what the whales sound like. I want to get inside their musical space, learn their style."

"If you imitate them, how can you tell if they're listening to you?"

"I want to learn from them."

"Me," he says slyly, "I'll try to raise my tune a half-step, see if they follow me. That's how I know they're listening."

"Sounds a little like science to me. Why should anyone copy anybody?" In music the interaction should be greater than the sum of its parts. Besides, I'd learned with birds that they have no sense of relative pitch, only absolute pitch. Raise your tone a half-step and it will sound totally different to them.

Jim is not pleased. "I've been doing this for years, David, and I can tell you one thing," he looks me in the eye. "Whales prefer a *groove*. They like reggae, they love the blues."

I protest: "You like reggae, you love a groove. I played in the Shy Five in Boston for years with those Bim Skala Bim cats, we once even opened for Pablo Moses. I dig reggae as much as the next eighties college student! But I'm not going to play it for the whales. You think I *want* to play these squeaks?—I know something about their hearing range, I'm thinking they'll like it, not me."

The next day we follow the winds. We're now on James O'Donnell's sailboat, planning to anchor at the original site of Orcananda, and to sit and wait. At night, the whales will pass by. No other boats will be around, the sea should be smooth, no grumble of engine noise. Then we'll see what they think of the clarinet and the electric mandolin.

O'Donnell has been a photography instructor, a salmon roe exporter, and a bush pilot. He tried to import Franco-Brazilian ultralight aircraft, only to watch a prototype explode in midair and kill two people. He runs a local seaplane airline that delivers mail and groceries to out-of-the-way homes in the archipelago, and now he's taking over another such airline. He hopes to sell both companies to the natives in about ten years.

James has been working with Jim Nollman and the orcas since the beginning. You'd think Jim N would be excited about our journey, but he still seems a little off. "Our interaction was of a period in history. It's come and gone. The whales that cared most about us have passed away. Today we pay them so much attention with all our noise, pollution, and whale watching gawkery that they would prefer to just be left alone. They don't want to sing with us anymore."

"Oh, I don't know," says James with the slow thoughtful drawl of a man who knows he can succeed at improbable things. "I'm one of the few people who have dived right down there with the orcas, looked them straight in the eye. They have always been able to sense our intentions, and are always ready for us if we come with sound purpose and mind. A group of humans can still focus the musical reach. Look how the orcas perform for every whale-watching boat! They *want* to be near us, they will want to make music with us if we approach them with the same wonder and desire as twenty years ago."

O'Donnell's crusty thirty-two-foot sloop is called the *Aquilla*, a fine name for the proprietor of Pacific Eagle Airways. I'm glad he's piloting, because neither Nollman nor I have any idea what we're doing on a boat.

Jim has managed to get the Beast fixed by a local electronics guy. What was the problem? We had plugged it in backward. I'm down in the hull listening to yesterday's recording. It is rather squeaky. To enjoy listening I have to transpose it all down at least an octave. The sounds of these northern resident orcas have been so well documented by John Ford that we ought to be able to compare what we hear to the lexicon of orca whistles I've got on my laptop.

Jim and James have bigger ideas. "I'd like to project a hologram into the air and have the whales be able to manipulate it," says O'Donnell.

"Yes," concurs Nollman. "I believe the whales transmit whole sound pictures to each other with their calls. Slow down a whistle or cry and it just becomes a bunch of clicks. It could be an audio image of echolocation."

An interesting possibility. To me the musical duet seems more immediate, and still unusual enough. Evolution produced music long before language. When whales gather, their cries might just be a celebration of the bonds that hold their clan together.

But there I am, sounding as New-Agey as the rest of them! What is it about whales that leads every person who sees them to feel an instant intimacy? We dream it into being. It's the same impulse that inspires all art and desire.

Over several hours we sail down the strait toward Orcananda as the sun sets. We see some whales and anchor the boat, ready to try again after some orcas squeal in the undersea distance. I play high clarinet squeaks along with their screams, and when I stop it seems they suddenly get louder. Interaction? Maybe. Nollman says no. He's after me again: "Your fifteen-watt sound system is a toy. There's no way those whales can hear you." His confidence is bolstered now that he's got his Beast working, streaming his electric mandolin out into the seas at a hundred watts or so.

"Too loud!" I bicker, a wind player's years of impatience with electric guitarists coming through.

"There's no way I could play softer," beams Jim. "Have to be twice as loud for those distant orcas to hear anything."

"I don't know about that, Jim. Roger Payne writes that even ocean-rumbling blue whale sounds are only ten watts in power. Even quiet sounds can travel several miles under water because of the way water carries sound." It's cold on deck and I warm my hands by rubbing them together. It's now nearly dark.

Below deck I browse the ship's bookshelves. I'm surprised to find a copy of *Let Us Now Praise Famous Men*, James Agee's gushing paean to rural poverty in the American South, with gripping photos by Walker Evans—another excessive national adventure only appreciated years after it was written. On the final pages Agee and Evans run across some strange alien noise in the southern forest night:

*Fig. 4. Jim Nollman and his electric whale mandolin.*

> From these woods a good way out along the hill there now came a sound that was new to us.
>
> All the darkness in near range of the earth as far as we were able to hear was strung with noises that were all one noise, and to this we had become so accustomed that this new sound came out of silence, and left an even more powerful silence behind it, so that with each return it, and the ensuing silence, gave each other more and more value, like the exchanges of two mirrors laid face to face.
>
> Whereas we had been silent before, this sound immediately stiffened us into much more intense silence. Without exchange of word or glance we each received communication of a new opening of delight: but chiefly we now engaged in mutual listening and in analysis of what we heard, so strongly, that in all the body and in the whole range of mind and memory, each of us became all one hollowed and listening ear.

Although I doubt it is what those guys had in mind, this seems a good primer for how to make music with the sounds of an unfamiliar being like a killer whale. In the moment of mutual listening everyone there dissolves into the presence of the sound.

Jim and James think I'm too prosaic when it comes to these orcas, because I'm not really sure that humans and whales share a special bond. "Ah, that's just because you haven't felt it yourself," beams O'Donnell. "Someday you will." For me the meaning might just remain in the fact of music, a sonic collaboration that is more than either species could accomplish apart.

Nevertheless, Agee gives me pause to wonder as he goes on about this mysterious sound in the deep Alabama woods (they think it's howling foxes but it sounds more like barred owls to me):

> This calling continued, never repeating a pattern, and always with what seemed infallible art, for perhaps twenty minutes. It was thoroughly as if principals had been set up, enchanted, and left like dim sacks at one side of a stage as enormous as the steadfast tilted deck of the earth, and as if onto this stage,

> accompanied by the drizzling confabulation of nocturnal-pastoral music, two masked characters unforetold and perfectly irrelevant to the action, had with catlike aplomb and noiselessness stept and had sung, with sinister casualness, what at length turned out to have been the most significant, but most unfathomable, number in the show; and had then in perfect irony and silence withdrawn.

Effortless and intricate at once, the perfect duet always lies beyond explanation.

Perhaps the difference between Nollman and me is that he's a blues man, a country mandolin maven who believes the whales are happiest to sing along when they hear an old-timey human tune, loud and clear. I too like a beat, but I want to join in with a whale groove, a rhythm I've never heard before. "Oh," says Jim, "then you're doing it for *you*, not for the whales. You just want to steal their stuff. What have you got to offer them?"

I have more doubt than faith. Who knows if the whales have any interest in a strange clarinet sound suddenly appearing in their midst? I've been listening to their cries, the pure whoops high up there on the scale. I let out my own crystal call. It's a bit shrieky for the other humans on board the *Aquilla*, but after a few seconds there's a high warbling answer:

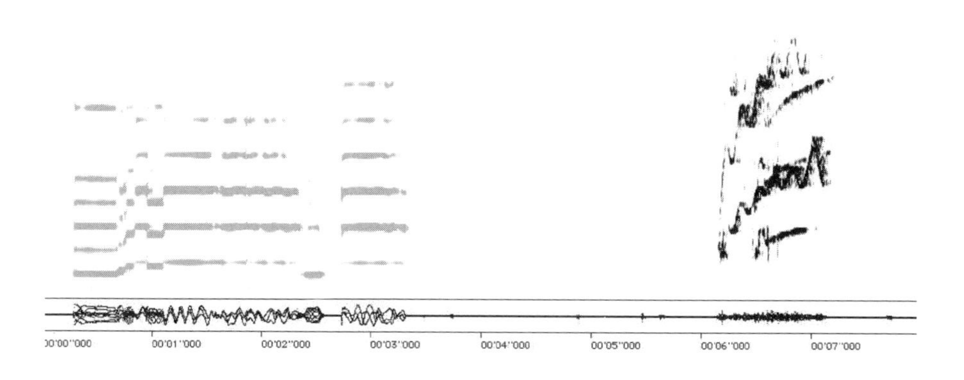


*Fig. 5. The clarinet plays a few sustained tones in gray, and the orca answers with warbling call N47.*

Those horizontal parallel lines in the gray clarinet sonogram in figure 5 show that I'm playing clear, sustained pitches. The rapid whale squiggles show the whale is making an amazingly weird one-second *whoeeep*, like a superfast slide whistle transposed into a squeak.

Many orca whistles are closer in form to the clarinet phrase I played, with simpler, steady tones. But that whale response looked unusual to me. When I sat down to pore through all the killer whale papers I had collected, I was amazed to learn that every single sound made by every orca in the northern resident population in the Johnstone Strait had been cataloged, and this is one sound that looked like no other: it's called N47, and it's sung only by the fifteen members of pod A1. So we know the whale who responded to me has to be one of that crew.

Nowhere else in the world and with no other species of whale have we so much detailed knowledge of the sounds the animals make and what kin relationships these calls reveal. When John Ford, as a young grad student in 1978, proposed that orcas might have different dialects in different pods, in order to tell one family from another, his professors laughed at him: mammals weren't supposed use sound in that way; it had never been observed in any terrestrial species. But they humored him, and let him develop his hunch into a master's thesis.

Hydrophones in those days were heavy, unwieldy devices, attached to thick insulated cables. Once he yanked one right out of the jaws of a curious orca, and the whale followed the microphone up to the surface, clicking loud echolocation beats onto the tape recording. The whale crested next to the boat and blew spray right in Ford's face. "It was enough to make me want to study mountain beavers instead."

Despite such setbacks, Ford was able to listen to whales for many hundreds of hours, and he soon realized his hypothesis was correct: each pod, a group of five to twenty animals, has their own characteristic repertoire. Over the years, he has been able to compile a massive catalog of every known Vancouver Island orca sound, focusing on the 250-odd northern resident population that ranges around the upper tip of the island, keeping out of the dangerous Georgia Narrows at the south end of the strait, which effectively separates the northern from the southern resident populations. Down in Puget Sound is the southern resident population, carefully monitored by Ken Balcomb and others. There are only about seventy-five of those animals, and they are also known by name and lineage.

The sound defines the pod, but not the individual. Yet the fin shape and coloration of orcas is often distinctive enough to tell one particular whale from another, so a photo atlas, like a family scrap-book or genealogy chart, has been compiled as well. This book has been published in two editions, the first in 1994 and the second in 2000, showing in fascinating detail how the population has changed. While the photos can help us differentiate each animal, they show nothing of how they might be related to one another. That information comes solely from their sound.

With years of data collected, Ford was able to demonstrate that although each pod had a distinctive repertoire, some pods shared certain sounds. Those with related but different sounds were then said to come from the same clan. It is believed that the whales tend to find mates from outside their pods, but inside the same clan. Since the pods are organized matrilineally, we only know for sure about the relationships between children and their mothers. The sounds are not genetically determined, but are taught by a mother to her children. But what is she teaching them?

After years of identifying and cataloging, Ford basically had this to say by the late 1980s: "No call type was correlated exclusively with any behavior or circumstance that could be identified." He surmised that the discrete, most identifiable calls help strengthen the bond between individuals in the pod and "coordinate activities." By the year 2000 he had a bit more to say, but not much: certain calls were more common between whales in the same pod, while others were more often heard when different pods met. In 2006 Paul Spong and his wife Helen Symonds published a paper carefully showing how certain family-specific calls were much more frequent following the birth of a calf. They suggest that these sounds, shared just in an immediate family group or "sub-pod," help to define an "acoustic family badge" in the first months of a baby whale's life.

In recent years Ford has focused on the fact that there are definite variations in how each pod-specific sound is performed, variations in inflection, pitch, energy. Ford believes these variations "appear to carry information on the emotional state of vocalizing individuals." Killer whales have feelings too.

And scientists are starting to believe that they have culture as well. Just as different pods and clans learn different sounds, different groups of whales have different behaviors as well, shared by all the members of each group. The most studied groups of orcas are these two resident populations, northern and southern. They mainly eat fish. But there are also other groups that pass through British Columbian waters. The transient whales are more worthy of the name "killer": they eat marine mammals instead, most often harbor seals, sometimes porpoises, and even the occasional moose! They hunt in packs and are able to stealthily approach other swimming mammals and tear them to shreds in minutes. They make far fewer sounds than the residents, because their warm-blooded prey have much more acute hearing than fish. It is believed these different behaviors are not genetically inherited, but passed from generation to generation as part of a tradition.

A third population of orcas lives in the deeper seas off the British Columbian coast. These are called the "offshore whales." They are rarely seen, little is known about them, but we do know they are a distinct population because they have their own distinctive sounds.

These three groups interact with each other but maintain separate repertoires, even though they meet in the same waters. Their culture is more complex than that of songbirds, where each dialect is usually characteristic of a specific area. The orca sound repertoire is not genetic, because the killers tend to mate with individuals from outside their clan, but still with members of their resident community. So the sounds must be learned, along with other behaviors that distinguish one pod from the next. Since behavior is passed down from mothers to children, the ways of life of killer whales are considered to be a form of culture.

Killer whales are found all over the world. Populations in Europe, New Zealand, and off of Kamchatka have been studied but nowhere near as extensively as around Vancouver Island. Worldwide, other populations don't behave in exactly the same way as the Canadian Pacific orcas. Ingrid Visser reports that in New Zealand the pods are nowhere near as distinct, as the whales seem to intermingle in much smaller groups. Those whales have been observed eating not only dolphins, porpoises, and fish, but also other whales and even penguins. There isn't the same distinction between fish-eating and mammal-eating groups—whale culture is different down under.

The sounds of the Vancouver Island population have been observed to evolve subtly over time, though nothing like the rapid change heard in humpback songs. Over the past thirteen years Ford and his student Volcker Deecke have closely monitored the sounds in several pods. Certain parts of the sound seem to change without rhyme or reason, similar to the process in human languages called "cultural drift," which explains why we sound different today from English speakers in the nineteenth century.

Since humans and orcas share so many aspects of group behavioral differences and group learning, it is not surprising that people and killer whales have a special relationship. Every one of those Canadian resident orcas has a name and number. You can even buy a deck of trading cards with each whale getting his or her own card! Orcas are the largest whales to have been successfully held in captivity, plus they made it big in Hollywood.

In 1993, the movie *Free Willy* opened to enthusiastic crowds of American children, describing the heroic efforts of a young boy, Jesse, and his Indian mentor Randolph to free a captive killer whale named Willy before the evil whale hunters could lure him back into his aquarium penitentiary. In the final climactic scene the whale leaps seventy feet in the air to clear a stone jetty and escape into Puget Sound. No real orca could manage that, and indeed it is an animatronic whale-robot who does the job.

However, there was a real whale in the film (only in the first film though, not in *Free Willy 2* or *Free Willy 3*). Just before the ending credits, an appeal came up on the screen: "You can personally help save the whales of the world by calling 1-800-4-WHALES." Warner Brothers expected about twenty-five thousand calls, but instead they got three hundred thousand. And more than half of them weren't about saving the whales in general, but about Willy in particular: What about that whale? Was he free? Can we set him loose?

Originally from Iceland, at the time of his Hollywood debut Keiko was languishing in a struggling, substandard aquarium in Mexico. He didn't get his fifteen minutes of fame from the movie, but from the overwhelming public response that followed. There was so much interest in the plight of Keiko from the world's children and their guilt-ridden parents that a special pen was constructed for him in Oregon, at a cost of more than $7 million, merely a temporary home, so he could eventually be reintroduced to the wild—something that had never been done with a killer whale. Next, an offshore operation in Iceland was set up, aimed at trying to resocialize Keiko so that he could return to his clan once more, decades after being sold into slavery for the pleasure of human children. Now in a pen in the open ocean, Keiko was surrounded by pods of wild killer whales, believed to be members of the clan he came from. Tens of millions of dollars were spent on Keiko's rehabilitation, more money than has ever been spent on any other single animal.

Keiko was attended by masseuses, tax attorneys, PR flacks, physicians, and animal psychics, some of whom said the whale directly spoke to them from thousands of miles away, saying he wanted to be free. Other channelers said Keiko told them personally that he wanted to stay in captivity, among the humans that loved him so. Nearly every child who saw him felt a personal connection to this giant whale.

Keiko was eventually let loose from his Icelandic pen, but he showed little interest in his relatives. Instead, he swam a thousand miles east to the waters of coastal Norway, one of few countries that still permits the killing of whales. But they don't kill killers, and Keiko loved people, so he swam up to our kind whenever he could. He died a few years later of natural causes.

Wouldn't those vast sums of money have been better spent on more widely applicable whale research aimed at conservation goals? Of course, but people are much more drawn to a cause when there is a single, struggling face behind it. That may be our biggest problem when it comes to dealing with animals: we care more about the one than about the many.

Kenneth Brower wrote the most complete account of Keiko's story, and he finds much to admire in this misunderstood whale: "He would leave sentimentality behind, and symbolism, and strained metaphor, and literary allusion, and Hollywood simplification, and the fantastic, anatomically incorrect poster-paint portraits of himself by his armies of devoted third-graders. He would escape the magical thinking of his channelers, the overprotectiveness of his trainers, and the righteous indignation of his advocates in the animal rights movement. . . . Keiko would outdistance every sort of falsehood, as those are all human inventions. He would course onward into bracing cold subpolar waters where everything is true. He would show us his flukes one last time and be gone."

Whales who have a special connection to humans get into all kinds of trouble. The 1970s science fiction film *Phase IV*, which is about two scientists trapped in a geodesic dome in the New Mexico desert, battling mutant ants who have developed intelligence and a mission to destroy humanity. But whales get in there too. The older scientist speaks to the younger: "I understand you did your dissertation on the language of killer whales. Now did you actually manage to talk with any of them?"

"Only the emotionally disturbed ones," answered the younger one.

"How did you know that?"

"Well, we talked."

I thought that was just a usual punch line until I heard the recent story of Luna, an orca who was sighted all over the shores of Vancouver Island for several years. Rejected by his mother at a young age, he quickly gained a fondness for humans and our boats, swimming very close to them, always getting our attention. Scientists said he was rejected by his pod. The natives said he was one of their great chiefs come back as a whale. He had an odd repertoire of sounds, including those of sea lions, providing the first conclusive proof that a wild orca could imitate another species' sounds.

Everyone knew this was a dangerous situation, that no orca could survive who loved motorboat engines this much. Some said his only hope was to live in an aquarium, but the current popular opinion on the island is that keeping any cetaceans in captivity is cruel. Others fought for a large, special outdoor enclosure where Luna could swim peacefully and relatively free, but there was no money to set that up, nothing like the venture capital that fueled the *Free Willy* circus. Whale watchers were of course cautioned to steer clear of Luna, but Luna would come right up to them. A woman was reputedly thrown in jail for patting Luna on the head.

Eventually Luna was sliced up by a motorboat engine. The captain didn't even see him. No one prosecuted the captain; everyone expected something like that to happen. "That's a lot for him to live with," sighed James O'Donnell.

One thing everyone did agree on in the end: Luna was an emotionally disturbed whale, who preferred the company of humans to his own kind. And we did not fully reciprocate by protecting him. He gave his life for this misplaced love.

O'Donnell tells another story of a man famous for diving with the whales and getting amazing close-up underwater footage of them. One day his equipment failed just a few feet below the surface and he floated up, dead. Blood had spurted all over the inside of his mask. The verdict was a faulty regulator. "His widow, though, had a whole different story," Jim solemnly continues. She said her husband had been talking for a few days before about being tired of living as a human, and thought about what it would be like to come back on this planet as a whale. "I think an orca zapped him with high-intensity sound. We know they can kill their prey like that."

A few weeks later the pod swam by her house, and with them was a newborn baby whale. The woman was reading *The Tibetan Book of the Dead* at the time. She still lives in the same spot, many years later, keeping her suspicions to herself.

"So, James," I ask O'Donnell. "When did you feel closest to these whales?" He smiles. It's an easy question for him to answer.

James and his ex-wife Muffin worked as observers and caretakers for Paul Spong in the early days of his whale research. They would sit by the shore and just watch for whales, day after day, fog or clear, taking note of what they saw.

One evening James felt a little out of sorts, and went down by the shore to meditate. He had been preoccupied for weeks with one simple question for the whales: "Orca, what do you want from us, what do you believe our species can share?" At that moment, four big whales surfaced with heads up, crowded around a young animal who was bleeding, injured. They were right near the shore.

O'Donnell felt a chill and a sudden message inside. "Human, is that what you want? You want to know what we whales and humans have to speak about? This is what we have to speak about." The adult whales pushed the injured animal toward him, whimpering, crying, now above, not under, the water.

His soul seemed to wrench apart inside him. The whales slunk under the water, and the surface of the sea was again made calm.

The next day Muffin heard on the radio that someone had shot a baby orca the day before, just for target practice. A few days later the carcass washed up on a nearby beach.

Not a man prone to flights of fancy, James has felt ever since that the whales have a clairvoyant sense that is able to reach out to us. His view is shared by many residents of the islands at the northern end of Johnstone Strait. After years of diligence, scientists have just scratched the surface of the animals' complex social lives and intricate sonic behavior. Meanwhile the inhabitants of Port McNeill, Sointula, and Alert Bay believe there are many mysteries about the orcas that no one who breezes in for a few days now and again can be fit to judge. We lost James in 2020. Rest in Peace.

More and more, people realize it is most important to observe these animals in the wild. The science of studying killer whales is meticulous and laborious, and the more we learn, the more complex their lives seem to be. The science of whale culture began with empathy, celebration, and experiment. By now it's gotten much more precise. How much do we honestly know about these whales and their world? Can the rigor of science outstrip the astonishment at hearing a whale cry for the very first time? The close link between human and whale that people who live up here enjoy and accept—that is a different kind of connection. It only comes with a long time spent rooted in the place. Their stories have a resonance that cannot easily be dismissed.

This is where interspecies music, rare as it is, might help. It's not supposed to sound like any tune or style that you know, but if you find a way into it, you may reach beyond the human vista and learn just a bit of what it must be like to be a whale. You're part of a group who all know the same sounds. Each of you interprets the family sound just a little bit differently. You sing to hold your family together, bright melodies guide you as you move. Your environment has gotten noisier: one cruise ship plows through your home every hour in the summer, not to mention all the freighters and trawlers all year round.

The sea is thick, there is not much to see down there. You ping your location with buzzes and clicks. Your sound may do more—perhaps you can stun fish, perhaps you can change your tune when you hear a clarinet or guitar reverberating under the waves maybe miles away. But you must be careful: some of these above-water beings want to learn from you or jam with you, but most through history have wanted to kill you, or poison you, without even knowing what they were doing. An animal may live in the moment, but their culture does not let history forget.

We're down in the galley listening to my recording and poring over the sonogram. Nollman is focusing intently. "Well, maybe you got something there, Dave, but I still think you ought to try reggae," he mutters. "You should try playing a skankin' backbeat for pilot whales in the Canary Islands, *mon*. Hear them blackfish, get a pulse going, and soon the whole pod will appear. Leave space between the beats, and it's uncanny how the whales fit in. Reggae is very complex rhythmically."

"Can't you play it just fine if you're stoned?" I remembered well enough. "Just ask Sly and Robbie."

Jim turns with a steely gaze. "It's difficult enough for human musicians to get it right. When I hear things like that I know the whales are very creative beings. Sometimes I think they must have a mathematical grasp of sound progressions to be able to pick up stuff so well. The ultimate goal is to get the whales to join the band." He looks out the window, as if gazing to the other side of the world, and back to those more perfect times. "In the Canaries I was playing off the Kairos, the famous New Age sex boat. Most of the people aboard were naked, getting it on with each other all over the deck. Me, I'm a happily married man. I was just playing my reggae to the whales, listening under water. Sometimes I had to close my eyes to concentrate."

Finally I can tell that Jim Nollman will never give up jamming with whales. He knows he's born to charge the border between the species. Since I went out with him to his favorite whale haunts many years ago, the Navy has even invited him back to help them sort out the conflict between sonar tests and whale strandings. The US government doesn't think he's crazy. And neither do I.

Back in the chilly autumn waters of the Johnstone Strait, things are much quieter. We understand the sounds of these orcas well enough to tell which clan each belongs to, who their mother is, and where they will go. But we're no closer to getting into their heads than we are at grasping what other people think, what music they like, what memories ring true. We all have our own private stories. Music is a way to sidestep these questions and dive right into the song, slipping easily into a beat, drowning the questions, listening for an alien voice that just might understand.

# 4

# TO HEAR THE DOLPHIN CALL HIS NAME

## How Smart Are They?

Animal intelligence is a hard thing to define, but if size and ratio are indicators, whales could be the most intelligent animals on the planet. They have the largest brains on Earth, with the most neuronal connections. Some species have the largest brain-to-body ratios of any animal. They are one of few creatures who possess vocal learning ability; like humans and songbirds, they learn to make their sounds. Cetaceans produce the longest, most involved vocalizations in the animal kingdom.

Whales and dolphins have complex social lives, where different groups may possess different cultures, evident in their different sounds, different ways to hunt food, and different forms of interaction. They may send sophisticated sound pictures to each other with a jumble of intricate sounds. Descended from animals who lived above ground, somehow they found their way back to the sea, where they never developed nuclear weapons or learned how to pollute.

Even though we inhabit separate worlds, the barriers between us and them can be softened. We can teach them to imitate our sounds, they can be trained to do interesting tricks. They can be taught to follow orders and aid us in wars they themselves would never want to fight. As they are roped into our world, we struggle to comprehend theirs. We used to hunt them and now we want to save them. We can swim with them and touch them. Or we can analyze them, collect data, ask questions. Which is the better way?

It's tough enough to define human intelligence, with the tests we devise for ourselves constantly under scrutiny. Certainly whales and dolphins are well able to solve the problems their undersea life demands. Inhabiting murky waters, they can see with clicks, find each other through sound, communicate for a thousand miles with low booming beats. And we appreciate them for this. We brag about their abilities. We're awed by their skills.

We humans have always admired animals who are interested in us. We admire what we value most in ourselves: individual curiosity, innovation, the ability to relate in a group. We had to train dolphins before we considered them smart. Dolphins follow our boats, sometimes save us from drowning.

Calling the alien sounds of whales and dolphins music may be no less anthropocentric than trying to give the animals intelligence tests.

Cetaceans use sound to detect the precise shape of objects in their environment, like a refined version of the sense some blind people possess.

Much more is known about how dolphins use sound than whales, since dolphins are smaller and easier to experiment upon in captivity. Toothed whales such as orcas, belugas, and even sperm whales are closely related to dolphins, so much of what we know about the sonic behavior of those animals stems from dolphin research. The baleen whales, with comb-like baleen for the harvesting of drifting plankton, are much harder to study. These include the humpback whale, with his magisterial song, and the giant blue and fin whales with their bellowing thumps that carry through the sea for a thousand miles.

In the 1950s science confirmed that dolphins use sonar to find their way in the water, detecting objects with a marine equivalent to the ultrasonic behavior of bats. The other whistle-like sounds they make help to identify individuals and build social cohesion among the group. Trainers in marine parks already knew that dolphins could be trained to do various tricks and to learn new sounds within the human vocal range—sounds they normally don't use—with the right incentives, usually food.

One man is most responsible for convincing the general public that these abilities are signs of high intelligence. John Lilly became the most popular writer on dolphins in his day because he built up people's expectations. He began as a neurophysiologist, figuring out how to use local anesthetic to attach electrodes to stimulate particular parts of the dolphin's brain. When he stimulated the pleasure center, the dolphin seemed to laugh. He was impressed at how well they could learn new sounds. This meant we could potentially teach dolphins our language and learn to communicate with them. He thought this would be perfect practice for learning to talk with aliens in outer space! "The extraterrestrials," he wrote, "are already here—in the sea." With his inspiring and sometimes extravagant claims, Lilly wrote best-selling books that created a generation of cetacean admirers.

Once dolphins were found to be able to imitate human sounds, we grew ever more fascinated with them. To Lilly this was a very strange moment: "The feeling of weirdness came on us as the sounds of this small whale seemed more and more to be forming words in our own language. The dim outlines of a Someone began to appear." Lilly's tone was much more exciting than the more modest claims of mainstream scientists. His work was supported by the National Science Foundation, the National Institute of Mental Health, the U.S. Air Force, the U.S. Navy, private foundations, and when he got interested in LSD and isolation tanks, reputedly even the CIA.

Even though Lilly inspired many to care about dolphins and whales, few cetologists today want to be associated with his name. He became a bad scientist. He would come up with wonderful hypotheses, but wouldn't follow through on them. While doing the experiments, he got restless and kept changing the rules.

But when most cetologists start talking loosely, they start to sound like Lilly as they wonder and dream. Lilly may not have been a careful researcher, but he was a maverick thinker. The biggest questions he raised still haven't been answered, and they continue to tantalize us.

The cover of Lilly's popular 1967 book, *The Mind of the Dolphin*, features a woman with a beatific smile embracing a thoughtful-looking dolphin. Margaret Howe lived with a dolphin named Peter in a specially designed house, half under water and half above. She was first there for a week, then a month, then for an extended period of *six months*, where she lived closer to the dolphin's world than any human has done before or since. Her own notes on these increasingly lengthy periods make for the most interesting pages in Lilly's book.

Margaret plays with Peter, swimming with him, tossing him his ball. "*Bawwww*," Peter learns to say. After many weeks of work, "*Maaahwgwit*." She sleeps in a wetsuit and spends her days seated at a half-submerged typewriter desk. The dolphin swims all around her. It's not easy: "To actually live with a dolphin twenty-four hours a day is a very taxing situation. . . . If given the opportunity, he will never leave your side. . . . To cross a room to answer a phone means that Peter meets you when you come into his immediate range and he walks with you, pushing, nibbling, slapping the whole way."

Peter gets a squeaky bunny. Peter has a square, a circle, a triangle. He learns which one is which. But he wants more. "I find that this living is hard and taxing on my private life. I do not think that I would like to live with this much restriction for too long a time." Think how Peter feels. Margaret can get up and leave the house but Peter has nowhere else to go. Yet the idea of isolation, of human and dolphin together, was key to the whole experiment. Live together, and you'll finally talk together. "*Bawwww. Maaahwgwit.*" But what does a male dolphin really need?

> Peter's sexual excitement usually begins with the biting business, and my stroking him. But after a few months, now, however, when his penis becomes erect, he no longer tries to run me down and knock me off my feet, rather he slides very smoothly along my legs, and I can very easily rub his penis with either my hand or my foot. Peter accepts either and seems to reach some sort of orgasm and relaxes. . . . Peter and I have done this with other people present. It is not a private thing, but it is a very precious thing. . . .
>
> I started out afraid of Peter's mouth, and afraid of Peter's sex. It had taken Peter about two months to teach me, and me about two months to learn, that I am free to involve myself completely with both. It is strange that for the one I must trust completely, Peter could bite me in two. *So he has taught me that I can trust him.* And in the other, he is putting complete trust in me by letting me handle his most delicate parts, *thus he shows me that he has trust in me*. . . . Looking back over the time spent and the notes collected, I find that I, for some reason, left out things about myself. Perhaps I felt they were not important or was ashamed of them.

What a thoughtful passage to have been written by a nineteen-year-old! And what careful self-awareness. I wondered where Margaret was today.

Shortly after reading *The Mind of the Dolphin*, I happened to be at a party in Woodstock at a recording studio on top of a mountain, and I mentioned to someone that I was researching a book on whales and sound. "My best friend's mother," she said, "once lived in a house with a dolphin."

"Oh," I said, surprised. "Was it on St. Thomas?"

"I think so. But she doesn't like to talk about it. Too many crazy people keep bothering her about those heady times."

"Maybe she'll talk to me. I'm not too crazy."

*Fig. 6. Margaret Howe Lovatt and Peter.*

I met Margaret Howe Lovatt at a vegan restaurant in the East Village. She still has that amazing smile, and her hair has only turned a tiny bit gray after forty years. In the early 1960s she had drifted down to the Virgin Islands after getting fed up with the advertising industry in New York. After working for a while at a hotel on the other side of the island, she heard about the dolphin research lab and decided to take a walk down there: "I heard there was this place on the east end, there were signs KEEP OUT! POSTED! WE DON'T WANT YOU! I kept walking all the way down the drive, and a woman came out scowling, 'What are you doing here? Go away!' She was very abrupt, but then this very tall gentleman came out, smoking a cigarette, and said, 'What's going on?' and I said, 'I heard there is a dolphin project here. I have some free time, I want to help.' And he said, 'We're just about to have lunch, come on in.'" That was Gregory Bateson.

Gregory Bateson—anthropologist, ethologist, ecologist of the mind—may have been the best collaborator Lilly ever had. He was a provocative and boundary-breaching scientist who tried to expand the way we think about our questions so they will lead not to easy answers but instead to ever better questions.

"We sat and talked for hours, and I had still never seen a dolphin," continued Margaret. "He gave me a yellow pad of paper and there was a spiral staircase going down to a pool and he said, 'Just sit there for an hour and write what you see.' I saw three dolphins. I had no idea what they were doing, and I just wrote and wrote, about five pages.

"Bateson sat and read it, saying ' *Hmmm.*' He laughed now and again, and he said, 'I'll tell you what, you can come here whenever you want. Just spend some time with the dolphins and watch.' Pretty soon I was part of the group."

Even though he was running the dolphin observation experiments at Lilly's behest, Bateson really didn't think what the animals were doing had much in common with language: "I don't think they're concerned with talking about things. I think they're concerned with the nuances of love and hate and respect. . . . It's going to be about *relationships*, not only to other dolphins, but to oceans, to geography, to navigation."

What was Bateson saying? Their sounds might be all about sound. Something not far from what contemporary dolphin scientists like

Vincent Janik believe, that the dolphins' use of whistles may be farther from language and closer to music, where meaning appears only in the way one sound is connected to the next.

Bateson saw that Margaret Howe had a real talent for the observation of animals, an essential prerequisite to success in ethology. "I didn't meet Lilly for a long time. Bateson was running this whole thing. Lilly was slick and polished, but Bateson was a very honest, creative, serious scientist who had an eye. And he liked my eye. He liked the way I saw things."

It was Margaret herself who came up with the idea of spending an extended period of time with a dolphin. By then Bateson had gotten himself a lifetime grant from the NIH and moved on to Hawaii to work at a marine park there (although he was never comfortable with the idea of training animals to behave according to a reward system, a practice that distorts their natural behavior).

"I told Lilly I would stay if I could do the experiment I wanted to do. Every night the dolphins were left alone, everybody goes home. This just didn't seem right. I wanted to live with the dolphins. I wanted to be in their environment, flooding a room, with each of us compromising, a bit uncomfortable—too much water for me, not enough water for the dolphin, but the two of us living together. Lilly thought this was a great idea!"

It was a way of learning he would never have had enough patience to do himself. "Lilly wanted things to happen, but I was fascinated when *nothing happened*. The most significant thing to me was to wonder why dolphins are so interested in us, in what we are doing. Remember: they don't really need things, that all comes from us. What do you talk about if you don't talk about things, if you don't have anything? My feeling is that birds *do* sing for no reason whatsoever. But I don't have that feeling with dolphins. I think dolphins move for the joy of it, they jump for the joy of it, but their sound has a purpose rather than just joy."

And years of study of dolphin sound-making in captivity has taught us a lot about that purpose. We know much about how they hear and how they produce sounds, but this information for large whales is much more hypothetical. Echolocation sounds are the ones most investigated, because of their sophistication and possible human applications. These high-frequency pulsed sounds emanate from the forehead of the dolphin in a precisely focused beam that is directed forward and slightly up. The dolphin adjusts the shape and size of various air sacs inside the fatty "melon" area, which functions as a kind of acoustic lens, focusing the sounds just before sending them out into the water. The melon is full of a rich, oily substance that is also found in the dolphin's lower jaw. Because of its role in sound production and reception, it is sometimes called "acoustic tissue." We do not know exactly how the melon's shape and form interacts with the sound signal to shape the beam of sound:

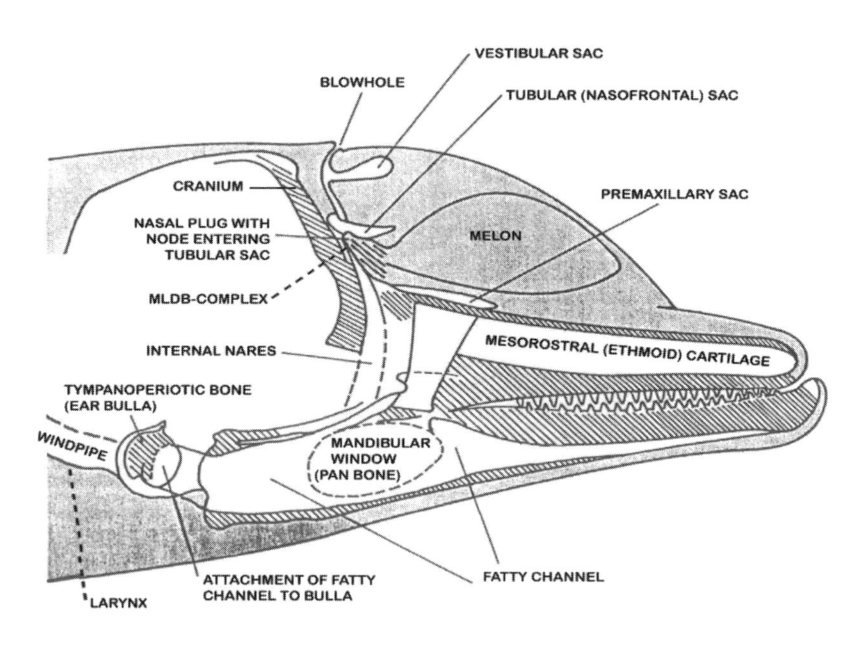


*Fig. 7. Inside the head of a bottlenose dolphin.*

It is believed that whistles are produced somewhere in the upper nasal passages, because air pressure inside these spaces rises just before whistle production and muscles around the nose are especially active as the sounds are made. But even though the whistles themselves are actively studied, the way they are produced has never been examined as much as echolocation clicks. Because whales and dolphins sometimes make several whistle-like sounds simultaneously, they may use different air passages at the same time.

In general we know very little about how cetaceans make their remarkable sounds. In the case of the expansive humpback whale song, we know nothing at all about how it is sung. Hearing is equally under-examined, but we know a little about how dolphins do it. Much of the listening happens through the acoustic tissue in the lower jaw:

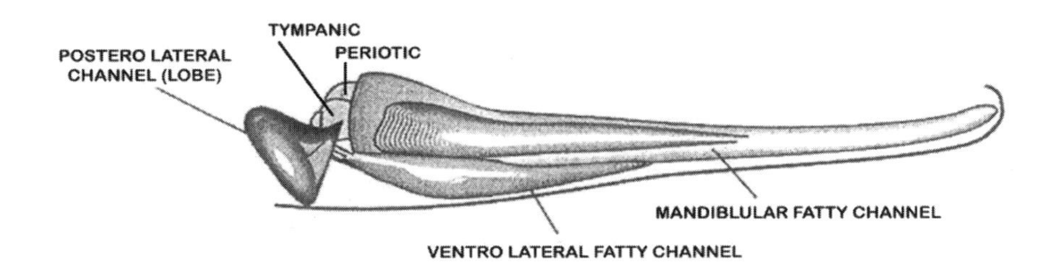


*Fig. 8. How the jaw of a dolphin contains organs that hear.*

By bouncing sounds off their surroundings, dolphins find their way through the opaque underwater environment using high clicking sounds, most far above the human range of hearing. Sometimes they are made so rapidly that we hear their low-frequency portion as buzzes and ratchety creaks. Decades of sonic experiments with dolphins have not taught us how they can precisely identify underwater objects primarily by sound. This ability is of deep interest to the Navy, who have learned much of how sonar can be refined by examining dolphin techniques.

In recent years Adam Pack and Louis Herman in Hawaii have determined that echolocating dolphins can accurately distinguish between objects like those pictured in figure 9 by the sounds they make and the echoes that return:

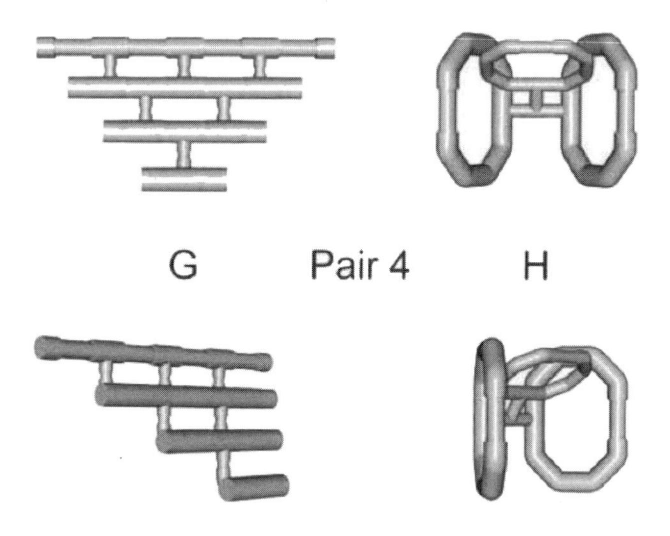


*Fig. 9. Objects that dolphins can distinguish between solely by echolocation.*

This experiment suggests that dolphins are able to store precise "sound pictures" of objects in their brains. They can positively identify such objects, even when perceived at different angles. The animals assess objects holistically, combining aural and visual senses, in a manner similar to humans. They may also be able to communicate these representations to other members of their species, but that may be wishful thinking.

Dolphins also have remarkable visual abilities. They can correlate the objects with images viewed on television screens. In fact, one of the signs of intelligence in dolphins has been that they like to watch television, which some might call a dubious achievement. No primates, dogs, or cats are especially interested in video images of themselves or their kind (even though there are special satellite channels designed for pets).

In order to be successful at echolocation, dolphins have to be able to discriminate the space between clicks with extreme acuity. It is believed they can detect intervals as small as .02 milliseconds between pulses, whereas humans can only go down as far as two milliseconds. They hear a hundred times faster than we do, when it comes to clicks. Birds, in contrast, can hear perhaps twice as fast as humans.

The other kind of sound that dolphins make is a tonal whistle, easily audible by human beings. It is believed that dolphins hear tonal sounds in a manner more similar to the way we do, although they are attuned to somewhat higher frequencies. Still, we can recognize clear distinctions among their whistles. In the 1960s dolphin trainers David and Melba Caldwell discovered that the animals produce a particular sound unique to each individual, which they named the "signature whistle." In captivity they produce this individualized sound most often when isolated from their mates and relatives, as if they are calling out their own name.

Baby dolphins make a whole range of whistles when born, but over the first half year of life, they develop and refine their own unique whistle. Female calves develop whistles very different from their mothers, but male babies tend to develop a whistle quite close to their mother's. This may be because the females spend a lot of time together as a group, while the males disperse much earlier. Signature whistles have been identified in both wild and captive dolphins, and they are most often heard when the animals are separated from their social groups.

Dolphins are also excellent at mimicry and imitation, which shows clear evidence for vocal learning ability, so rare in the animal world that it is shared only by humans, birds, and cetaceans. No chimp or other primate can manage such a feat. What sounds do dolphins choose to imitate? We know that Lilly and McVay were trying to teach them English words, but in the wild, they copy the signature sounds of other animals they know. "Although dolphins can play with imitating novel sounds," writes Woods Hole scientist Peter Tyack, "they harness learning primarily to develop stable whistles." This is exactly how humans learn words, playing around with our flexible sonic ability until we settle in on the sounds we need in each context. "If dolphins learn to develop their signature whistles and learn to imitate the whistles of others in order to name themselves, then vocal learning allows a remarkable open-ended system of communication in these social animals."

Tyack describes an experience in 1984 when five wild dolphins off Sarasota, Florida, were in the study corral, and one was in the "processing raft." The dolphin in the raft was named Nicklo, thirty-four years old. She whistled eighteen times a minute throughout her time in the raft, higher than any of the freely swimming animals, producing 520 signature whistles, 39 variant whistles, and no imitations during the first half hour. In the second half hour she began to imitate the signature whistle of Granny, the oldest dolphin in the corral, forty years old at the time. She imitated no other animal, not even her three-and-a-half-year-old calf. Perhaps, imagines Tyack, Nicklo acknowledged Granny's seniority in the group, and recognized that she might have had previous experience with this kind of strange captive situation, and might be able to help.

This kind of anecdotal evidence (you could call it "Lilly's Legacy") infuriates some scientists. They resent the fact that such stories spread quickly through the popular media, while the evidence to back them up is spotty. For such a conclusion to be truly rigorous, the experiment would have to be repeated many times and such behavior observed in a statistically significant majority of instances. Cetacean science is full of such tales, and it becomes very hard to separate the truth from wishful thinking. But the general public doesn't need all that evidence—a good story can spread quickly around the world.

In 1998 the great bird song scientist Peter Slater, along with his student Vincent Janik, pointed out that although signature whistles are easy to cajole out of a dolphin when you separate one from its group, we know little about how these sounds function in the usual life of the animals. The researchers monitored the sonic output of four captive dolphins, when they swam together and when one was separated in another pool. When separated, the lone animal produced only its own signature sound. When they were all together, that signature sound was never heard. Instead they all whistled a similar sound.

The paper concludes that, in captivity, signature sounds are used exclusively in situations of separation, confirming earlier suspicions. Slater and Janik also point out that with such a clear context for their use, signature whistles are much easier to study in dolphins than the sounds they share, which could only be investigated in much more complex experiments, like the kind tried on birds who share and match songs in a variety of exact contexts. But dolphins are much harder to observe in the wild.

Tyack has never been fazed by this difficulty. Together with his student Laela Sayigh, he tried to investigate how female dolphins respond to the whistles of their grown-up offspring who live independently. They found that, indeed, the mothers more often turned their head to the unique sound of their adult children than to similarly aged animals to whom they were not related. The same in reverse, the kids were more interested in the sounds of their mothers than other females of the previous generation. It is not a surprising conclusion, but it does suggest that dolphins are able to recognize individuals in their community through sound.

It is unusual for every individual in a group of animals to possess its own unique name sound. So there are those who doubt this theory. Brenda McCowan and Diana Reiss have long argued that the signature whistle hypothesis is just that, a hypothesis, one that too often colors the design of experiments as well as their results. If you believe in signature whistles, then you will design a situation that makes it look like they exist. You will focus on these whistles, which aren't produced most of the time, and ignore all the other whistle types made by dolphins, the sounds that they share together when in a group. McCowan and Reiss point out that the signature whistle *idea* is such an appealing one that it colors what scientists see and hear. Before the hypothesis was proposed by the Caldwells in the 1960s, some studies showed that dolphins isolated from their normal groups produced a similar characteristic rising whistle most of the time. After the hypothesis, everyone started to hear a distinct whistle unique for each animal.

McCowan and Reiss's attempts to replicate the experiments of Tyack, Sayigh, and Janik came up with very different results. They found that isolated lone captive dolphins produce a *shared whistle* most of the time, with only very subtle differences unique to each animal. This is a much more common vocal strategy in the animal world, well known in mammals like the squirrel monkey, and famous in penguins and other birds as well. Individuals are more likely to be recognized by subtle differences in their voice than by wholly unique whistles. The two women used a form of mathematical analysis to detect these slight differences in the sonograms of the whistles.

Their paper also notes that most dolphin whistles have not been the subject of serious investigation. If we get beyond the signature hypothesis, we may find that the way dolphins use whistles is far more complicated than just announcing their own name when they are in trouble.

When I asked Janik about McCowan and Reiss's conclusions, he criticized their methods: "They only paid attention to the whistles dolphins make when they produce bubble streams that rise to the surface. The animals mainly produce bubble streams when they sing upsweeping whistles, not other kinds. So those are the only kinds of whistles they noticed. Plus, they do not tell us *how long* they isolated their animals before recording their whistles. Some animals need to be alone for quite a while before they produce their signature whistle." And he didn't trust their algorithm, maintaining the human eye assessing a sonogram is still more accurate than the calculations of any machine.

Meanwhile, the signature whistle hypothesis garners ever more support. Peter Tyack simply told me, "In my experience, signature whistles are so obvious it is barely worth debating." An important paper on the topic was published in 2006, by Vincent Janik, Laela Sayigh, and Rebecca Wells. In response to McCowan and Reiss's view that the dolphins identify signature information in the subtle tones of the voices of animals they are closely related to instead of the pitch contour of the whistle being used, this team decided to play back not actual dolphin whistles from members of the test dolphin's family but artificial whistles produced by a computer program called Signal 3.1. It's a carefully designed experiment based on the approach taken by Erich Hoyt out on the water back in the seventies when he first played a synthesizer to see if orcas might respond to electronic sounds.

The synthesized whistles worked just as well as the real ones, demonstrating that the shape of the sound is what counts. Humans can identify who is speaking by tone of voice; the same word sounds different when spoken by different people. That's the way most animals tell who is vocalizing as well, but not the bottlenose dolphin. Janik's latest experiment shows that it is the contour of the whistle, not the timbre, that matters. It's the name itself, not the way the name is said. A dolphin, when separated from his family and friends, tends to speak his name more than any other sound. This is the strongest evidence for the signature whistle hypothesis thus far.

McCowan and Reiss still think everyone else is brainwashed by hopeful bias. "I would be delighted if there was strong supportive evidence for referential signature whistles in dolphins," assures McCowan. "It would be the first evidence for referential capability by dolphins in their own communication system. But I simply don't think the data currently support this interpretation." She says the actual curves of the supposedly distinct signature whistles are not different enough to be called discrete signatures. She's not impressed that the dolphins respond well to simple synthesized versions of their whistles. The rich tonal quality the animals are able to produce naturally suggests much more potential for complex communication. "That dolphins, using air sacs instead of their larynx, can modify their whistling into very different whistle contours and can even learn to imitate novel computer-generated whistles, as we and others have shown, speaks much about their ability to control their whistle output, perhaps much like humans."

In 2007 Sayigh and Janik responded in *Animal Behaviour* to McCowan's critique by reclassifying the original data using her method, and theirs. They maintain that their original analyses are correct, stressing that McCowan and Reiss used a classification method that is biased toward noting similar kinds of whistles, not whistles that come from the same animal. The two women did not listen long enough to each whistle, and focused too much on the presence of the upsweep, not realizing that this *whoop* could be a component of a distinct signature whistle, if they chose to listen long enough to the sound.

Most scientists working with dolphins today support the signature idea. Sylvia van der Woude in Berlin listened to the whistles in the dolphinarium in Eilat, Israel, for several years as part of her doctoral research at the Free University. She points out that she can easily distinguish individual dolphins by their signature whistles while listening to them in their open-sea pen, "and humans are not nearly as acoustically acute as dolphins." Mammals on land may well rely on innate vocal cues, but McCowan and Reiss have claimed the same approach works under water, where tonal quality is much less consistent, due to compression and deformation of the vocal tract while producing sounds.

How signature whistles relate to the rest of the dolphins' whistle repertoire is largely unstudied, and that is the subject of van der Woude's dissertation research. "When a dolphin imitates another animal's signature whistle, it is likely a calling out for the other animal. But how can the imitation be distinguished from the original?" Maybe McCowan and Reiss are onto something after all. "There may be subtle vocal cues that reveal the identity of the imitator," says van der Woude. "And yes, there are many more specific whistles the dolphins use besides their own signature, they are right about that."

Van der Woude reminds me that dolphins make all kinds of other sounds besides echolocation clicks and whistles. Besides burst pulse sounds, such as grunts, barks, brays, and screams, there are wheezes, gulps, and even bangs. She has found very low moans that sound a bit like humpback whales. No one has published much on the meanings of any of these cryptic noises. "Too much work on dolphins is based on training, not on what they naturally do." It's all biased toward the human realm, like Lilly and McVay's early attempts to teach the animals to talk.

Whistles may convey much more than the whistler's name. Humans have developed whistled languages all over the world, wherever we need sound to travel far. Most famous is the *silbo* language of the Isle of Gomera in the Canary Islands, an ancient whistled system that nearly died out in the twentieth century, but is now required study in the island's schools. People developed whistled languages in response to environmental constraints, as the behavior of dolphins has evolved in response to their habitat. Human whistles can be heard distinctly as far as three miles away, much farther than speech or shouting.

For whistling to convey the information contained in human speech, vowels are replaced by distinct ranges of pitch. Accent or stress is represented by slightly higher pitch or a stretched-out whistle, and consonants are represented by different speeds of transition from one pitch to another, or by different kinds of attack on the whistle, say, "*tooo*" versus "*ooooo*." In Silbo, five vowels and four consonants are produced, enough to convey much of the information in spoken Spanish. Whistled human languages, although fairly rare, have also been studied in Turkey and Mexico. Can these studies help us understand dolphins? Researchers mainly point to the fact that much complex semantic information *could* be contained in whistles, if the animals need to do it.

In the years since the first edition of this book, evidence for the signature whistle hypothesis seems to have increased. In 2013 Janik and Sayigh wrote a comprehensive review of fifty years of signature whistle research. Audra Ames wrote a fine master's thesis in 2016 investigating the whistle duets between mother and daughter dolphins, and in 2014 Stephanie King investigated the precise timing between dolphin whistle matching. Dolphin science marches on. And in Diana Reiss's fine book *The Dolphin in the Mirror*, she does concede, "We do know that among the rich and varied repertoire of whistles, squawks, and other types of calls the dolphins produce, the most frequently used signal is a contact call that conveys signature information about the caller."

A pod of dolphins whistling together remains a mess of unfamiliar sounds. Some scientists have even asked musicians to help them figure out what is going on. Doudou N'Diaye Rose is so important a musical figure in Senegal that no Air Senegal plane will fly a maiden voyage before he plays a special rhythm to ward off a crash. The late Dutch dolphin expert Cees Kamminga invited Rose to listen to a recording of dolphin whistles in the mid-1990s. Without learning anything of what the scientists suspected was happening, he immediately offered this interpretation of what he was hearing. "There are three animals on the recording. The dolphins first give their names, and then they begin a conversation, and they are able to share each other's name sounds." With one listen, he had immediately heard signature whistles at work, in a very sophisticated manner that was only hypothesized ten years later by Vincent Janik.

Musical acuity might help us understand cetacean sounds, if better experiments could be devised. Sadly, Kamminga died in 2002, and his work stopped. German composer and radio producer Michael Fahres was deeply touched by the story, and he was able to get funding from cultural agencies to get Doudou Rose and live dolphins together in person. Rose tapped rhythms on the glass walls of their aquarium and the animals swam close and listened. He played a bucket of water like a drum and the animals were transfixed. Fahres traveled down to Senegal and worked for several weeks with Rose's Master Drummers of Senegal ensemble, comprising a selection of his fifty sons. Fahres's composition "Cetacea" combines Western elements of conducting and structure with the traditional West African approach of rhythms passing instruction on to other rhythms. They all traveled to the Traumzeit Festival in Duisburg to premiere the work in the spring of 2006.

Six drummers sat at the edge of an open area. Flat onto the floor, a film was projected of swimming dolphins from an oceanarium. Prerecorded dolphin sounds, transposed down many octaves into the range of human hearing, swirled around a multichannel speaker system as Fahres mixed the live and taped elements together. The theory of overlapping cetacean rhythms was finally turned into interspecies art. Dolphins say their names, all together, and the result is not only data, but music as well.

We still don't know how smart dolphins are! For that there is no better person to turn to than University of Hawaii professor emeritus Louis Herman, known for the remarkable echolocation experiments described earlier. "I'm thoroughly impatient with speculation and wildness and extrapolation," says Herman. "In scientific life there's no substitute for hard work and patience." In 2006 he wrote the chapter on dolphins for a book entitled *Rational Animals.*

"A rational animal," writes Herman, "is one that can perceive and represent how its world is structured and functions. . . . The function of a mind is to create a model of the world." The animal can draw conclusions about this world in order to act more productively within it. The animal can learn new information and then modify its behavior accordingly. And most importantly, these abilities are based on an innate and species-determined level of intellectual capacity. The intelligent animal can go beyond the boundaries of the familiar.

The way to investigate this is not to take physical measurements of the animal's brain. It's far better to watch the animal at work in challenging situations. This is what Herman has been doing for decades. In the 1980s Herman taught dolphins to imitate many different sounds, some human, some cetacean, some electronically synthesized. Sometimes the animals could copy the new sound in minutes, and sometimes it took them many hearings over many weeks to get it. When faced with a sound above or below his preferred vocal range, the subject dolphin Akeakamai (Hawaiian for "lover of wisdom") would attempt to mimic the sound, but he did it an octave below the sound that was too high, and an octave above the sound that was too low. Although this phenomenon of "octave generalization" certainly seems like a good solution to the problem, it is extremely rare in the animal world. For all their impressive mimicry abilities, birds cannot recognize sounds an octave away as being the same note. They do not possess a sense of pitch relationships.

Anyone who has been to an aquarium knows that dolphins can be taught to perform elaborate routines to delight and amuse human audiences. Maybe it's all just to say "thanks for the fish," but Herman demonstrated that dolphins can go one step further. He taught his dolphins a sign that meant "create," instructing them to make something up; either their own routine combining tricks already learned, or else an entirely new behavior no human had ever instructed them to do. For example, Ake knew many behaviors, including *back dive*, *blow bubbles*, *tail wave*, *open mouth*, *spiral swim*, *spit water*, and *pirouette*. Tell him to "create," and he will put them all together in a new order, or add his own action to the mix.

Dolphins have also been shown to understand grammar, so that they know that the phrase in symbols "surfboard swimmer fetch" means bring the swimmer to the surfboard, while "swimmer surfboard fetch" means bring the surfboard to the swimmer. These are definitely levels of understanding that few animals have shown a similar ability to follow.

Why might cetaceans have evolved such abilities that few other species share? Herman suggests that the animals' complex social lives require highly developed mental abilities. This is a more significant indicator for him than the need to process echolocation clicks to find their way under water. Why does life under the sea require that the animals need to communicate with each other so effectively? Perhaps it's as simple as the fact that dolphins have little protection against shark attacks, save each other.

After all, some scientists say the need for *warfare* is what required that humans learn to talk to one another. We had to cooperate against the enemy. Surely there is more to it than that. The fact that dolphins can be trained so well has made them of great interest to the military. They may even be able to stun their prey with loud and sudden sounds. This was considered hearsay until 2001, when Ken Marten and Denise Herzing videotaped a dolphin emitting a loud "*pow pow pow*" sound while chasing a fish.

These facts about dolphins seem to be greater than any possible fiction. "Because we can't find out very much about dolphins," said pioneer researcher Melba Caldwell, "mystery does remain. Then you can attach a myth to a mystery. But the thing is: do you want answers to your questions, or do you want to keep your myth?" Science's answers may be much less satisfying than our dreams.

The 1974 Mike Nichols film *The Day of the Dolphin* is a thriller whose plot involves a dolphin scientist whose research is co-opted by the military for immoral plans. George C. Scott plays scientist Jake Terrell, clearly based on John Lilly, with a laboratory on a small Caribbean island where he is conducting experiments on dolphins that he doesn't want to talk too much about. In the initial scene he's answering questions from an audience on how much the military is involved in his research. "Oh no," he answers, "the Navy doesn't talk to people like me." In the next scene Scott is back on the island, speaking with his beautiful wife and assistant, played by Trish Van Devere (who was Scott's wife in real life). "Was Alpha being aggressive with you in the pool again?" "Oh yes dear," she feigns embarrassment. "Well, we simply must get him a mate."

A female dolphin named Beta, nicknamed Bea, is brought to Alpha's pool, and they spend the days playing side by side. Only after Bea is removed to a holding pool, and Alpha can't be next to her, does she finally learn to speak to her human "Pa." Alpha, who knows himself as "Fa," says, "Fa Need Bea *now*!" Through sex and necessity, interspecies communication begins.

From there the plot runs wild to a conflict between the scientists and their funders, a sinister foundation that wants to use the dolphins to blow up the boat of the president of the United States. Dr. Terrell must teach Fa and Bea that "Man Not Good," or "Some Man Not Good." The animals figure it out and blow up the boats with the bad guys. Then they are told to swim away forever. "Fa Loves Pa," cries the dolphin. He does not want to go. But as the film ends, more of the bad guys are coming to the island for the scientists. The dolphins disappear out to sea.

The comedian Buck Henry, who wrote the screenplay based on the French novel *Un animal doué de raison* [*A Sentient Animal*] by Robert Merle (quite a different story, in which the dolphins carry on full-blown conversations on important issues of the day), did the dolphin sounds himself. "After doing those voices I couldn't get work being any other kind of animal, I was pegged as a dolphin man," he says on the DVD commentary track. Also remarkable about the film is how director Mike Nichols, most known for *The Graduate*, knew from the beginning that he wanted to use wild dolphins, not acclimated aquarium professionals. So he had two dolphins caught and trained just for the film. When they were done filming, they were released into the wild. Henry remembers the wrap: "After we shot the last takes of the dolphins saying good-bye we just let them swim away, never to return. I wish more of our actors behaved like that."

*The Day of the Dolphin* played well upon the public's hopes and fears about what scientists and the government were doing with dolphins. What were they actually doing? Journalist David Helvarg reports that today, the Navy spends nearly $20 million a year on the study of more than a hundred marine mammals at a secret facility at the Point Loma submarine base near San Diego. In the early days of the Iraq War, in 2003, the Navy's trained dolphins and sea lions detected a number of undersea mines, and also searched the waters for enemy swimmers, using their unparalleled natural sonar abilities.

The military assures us that there are not now nor were there ever dolphins trained to do anything more aggressive than this. Helvarg found retired Navy trainers with a different tale to tell. Former CIA dolphin specialist Michael Greenwood admitted that in 1971 six Navy dolphins were airlifted from Hawaii to Cam Ranh Bay in Vietnam, where Vietcong divers were attacking American ships. The dolphins were fitted with padded cones on their beaks containing a hollow needle attached to a carbon dioxide cartridge, a device originally designed to be used in a shark attack. The dolphin would swim to an enemy diver and plunge the needle right into him, exploding his insides in an instant. This was called "swimmer nullification."

The Navy asserts that dolphins "move so quickly and with such accuracy that human swimmers in dark or murky waters are located and marked before they know what has happened." But why stop with marking the bad guys when they could be eliminated? Media reports in the late 1970s claimed that Navy dolphins killed several dozen North Vietnamese soldiers, and two Americans by mistake. The dolphins were also known to have attached mines to undersea piers so they could be blown up. Several former trainers of these animals told Helvarg they left the program because they questioned the reliability of these cetacean operatives, as well as the ethics behind the whole thing. "You don't see the Air Force supplementing its predator drones with camera-wielding bald eagles," says Helvarg. "War is essentially a human activity that we really don't need to share with our fellow mammals." That ethical view applies to whatever level of war games the dolphins were taught to play.

Today dolphins and sea lions are being trained in San Diego to guard the Kitsap-Bangor Naval Base in the state of Washington. They are still used to locate underwater mines and help rescue struggling swimmers. The Navy maintains that the animals are still far more sophisticated than any technology humans have been able to create. The Russian navy has made use of dolphin and belugas in the war in Ukraine. Rational animals indeed! Will we let them choose which side to be on?

I remember a table in John Lilly's book *Man and Dolphin* that still makes me laugh. Humans in the left column: wage constant war against each other, killing millions of our kind. Dolphins in the right column: live peacefully in the sea with a minimum of bloodshed. Who shall we call more intelligent? I wouldn't call this chart scientific, but it's still a good question.

Science essayist Loren Eiseley explained our distance from their world even more poignantly, "No matter how well we communicate with our fellows through the water medium we will never build drowned empires in the coral; we will never inscribe on palace walls the victorious boasts of porpoise kings. We will know only water and the wastes of water beyond the power of man to describe."

We praise them for being able to call out their own names, and applaud the complex tricks they can perform. We can drum along with them, and even dare to live close to them, half out of the water and half in. Yet we will never feel what it's like to live confined by the sea and be able to sketch pictures of that world in sound.

# 5

# BELUGA DO NOT BELIEVE IN TEARS

## Russian Whale Music

Beluga whales live in all the world's Arctic seas. They are graceful and beautiful, enjoyed by everyone who meets them. Sailors called them "sea canaries" because they make so many sounds. They are kept in aquariums all over the world, but Russia is the best place to meet them musically in the wild.

"You should come with us to find the white whales of the White Sea," says Rauno Lauhakangas, the Finnish physicist, when I reach him in the middle of a particularly tricky experiment in his laboratory. (Jim Nollman had given me his name.) "No government will bother us there. Next summer we're going to have divers wearing white suits. We think white is an important color for the belugas, and they might like us more if we're wearing white. And we hope to set up a directional hydrophone array so we can really hear which one is making which sounds." He has already heard of my musical interests. "Playing music with them is fine. There are exactly seven spots in the White Sea where the belugas definitely congregate in the summer months. Next summer we're trying out a new one, the island of Myagostrov, in the Republic of Karelia—What, you haven't heard of it?—Be prepared for a journey back in time of at least one hundred years."

At least we'd be far enough from anyone who might accuse us of breaking any laws if we tried to use music to communicate with a few white whales. In America and Canada the animals are protected by the Marine Mammal Protection Act, and anything one wants to do with them is carefully reviewed by a panel of scientists before it is approved. The Inuit are allowed to hunt them in their territories of Nunavut and Nunavik in Canada, but even they are not allowed to play tunes to them. Music making is no longer considered research; with whales, science trumps art.

The beluga whale, whose name means "the white one" in Russian, might be one of the best species to try and make music with. Their wide range of whistles, clicks, and buzzes is far more diverse than the dolphins, whose sounds and behavior have been studied the most. In 1585 the Dutch traveler Adriaen Coenen wrote that their "voice sounds like the sighing of humans. . . . If a storm is imminent they play on the surface of the water and they are said to lament when they are caught. . . . They like to hear music played on the lute, harp, flute, and similar instruments." Even in the sixteenth century people played concerts for the whales! For a long time people have sensed that these animals are intrigued enough by human life to enjoy listening to our songs.

There are many places where belugas are successfully kept in aquariums and marine parks, from Vancouver, Mystic, and Atlanta, to Valencia in Spain and Kanagawa in Japan. It was the Shedd Aquarium in Chicago that offered the most encouraging response when I asked them if I might play some music to their whales. Ken Ramirez, chief animal trainer and vice president of marine mammals, wrote me back a detailed series of questions to test the nature of my interest. *Why belugas?* I answered immediately: because they are known for their vocal ability and interactivity. *How will you be communicating with them?* Play the clarinet into a microphone, plug the microphone into an amplifier, the amp into an underwater speaker, dunk the speaker down into the pool and let the whales hear it. *How will you record the interaction?* Stick a hydrophone into the water and feed the results into a digital recorder.

Ramirez offered to let me spend three mornings among the whales, and he would accompany me the whole time. The Shedd was not the first aquarium I had contacted. The Mystic Aquarium in Connecticut initially gave me permission but then rescinded it, an order sent down straight from the director. I never did find out exactly why—they told me it had something to do with secret research their belugas were going to be part of, probably something highly classified.

It may also have had to do with the fact that there is deep disagreement among whale researchers and whale aficionados as to whether these animals should be kept in captivity at all. They may have thought I was an activist for beluga freedom. In the wild, white whales routinely travel hundreds, even thousands of miles through open water. Is keeping them in an exhibition pool the equivalent of prison? Zoos and aquariums alike have had to deal with such criticism in recent years, and there are several publicized cases of former dolphin trainers and whale scientists who have then turned their attention to the release of these animals, from Paul Spong to Rick O'Barry. Indeed, after the initial publication of this book, an offer for me to speak in Scotland was rescinded after the organizer of the event decided I was not sufficiently critical of the keeping of cetaceans in captivity.

Yet the animal keepers and trainers I have met in aquariums and aviaries often have much more intimate knowledge of their charges than those who observe elusive creatures in the wild. They would have to, because in captivity animals depend on our attention to survive. Many of the details of animal behavior and physiology we have discovered could only be learned in the close confines of zoos and aquariums. Most people learn of whales from books, film, television, and recordings, but aquariums give many of us our first chance to see these animals in person, an experience deeper than any media image. That being said, the tide has certainly turned against keeping such intelligent animals in tiny pools for the benefit of our own amusement. Facilities like the Sea World Park who keep orcas in captivity are steadily losing support after hard-hitting documentaries like *The Cove* and *Blackfish*, proof that clear-cut storytelling can really sway public opinion. Of course all this interspecies work is better in the wild, and over time, we are learning the patience necessary to listen and engage out there.

But for Ken Ramirez, the chance for direct experience is the most persuasive argument in defense of his facility. He is an inquisitive and dedicated caretaker of belugas and dolphins. While president of the International Association of Marine Animal Trainers, he assembled a seven hundred-page book on incentive training. "I work with these animals every day, I feel I *know* them. It's my job to make sure they are content. All of our whales are born in captivity," he assures me. "It is doubtful they would find it easy to survive in the wild."

One of the fundamental principles of his care for the belugas is *enrichment*, the idea that the animals' lives can be improved by giving them new things to think about, to consider, to engage with: new toys, new tests, new games to play. Ramirez saw my interest in playing music to the belugas as a potentially enriching activity, as I would offer new sounds for them to consider. "When you bring an animal into an artificial environment, you are not replicating the wild. There is no substitute for the ocean. Part of good animal care is thinking about their health, nutrition, environment, and behavior. And we provide substitutes for what they're missing. In the wild, animals are very active, because they're always hunting for food and avoiding predators. We need to create activities that keep them physically fit and mentally stimulated."

Ramirez and his colleagues try to enrich the lives of their belugas by offering them options that make life in captivity more fun. The whales can choose to swim from one pool to the next. If they want to play with a toy, they can. It's up to them. He's tried music many times before: "When we play music we open up gates, and allow the animals to move. Do they come close, or do they move away? We can find out whether they are attracted to particular sounds, or repelled by them. We observe what happens, and that's how we learn."

The Shedd Aquarium is a beautiful facility right on the shore of Lake Michigan, and the beluga habitat is its centerpiece. With mock evergreens and plastic rocks, we're supposed to feel we are on the coast of Alaska or somewhere north where these astonishingly white whales cavort in the wild. Here they migrate only from one observation pool to the next, rarely standing still. How does Ramirez know they are happy? "Just as with dogs and cats, when you spend enough time with particular animals, you learn what they need. An animal that is attentive and engaged, she's happy. One that sits silently in a corner, barely moving, is not." The sounds of piped-in loons and imaginary waves echo around us, and if you look out the giant latticed windows to the gray-green surface of Lake Michigan, it is not hard to imagine that you are somewhere real and outdoors, maybe even the White Sea.

I imagine a beluga leaping out of his tank, sliding over the rocks and disappearing into the great, gray lake. How would he like it out there? He could swim through the locks to his compatriots in the St. Lawrence estuary, the most polluted population of belugas in the world, isolated from the rest of their kind who move through much wider swaths of sea far to the north. These aquarium whales are probably much healthier than those in that vast estuary a thousand miles to the east. Who is better off? What is captivity to these animals who have known no other life and bring such joy and amazement to all the parents and children who visit them?

Some people believe no cetaceans should be kept in captivity, arguing that zoos by their very nature are a cruel form of prison. Ramirez has heard such talk for years, and he does not agree: "This profession has come a long way." The first captive dolphins and whales were kept purely to put on a show. But today they are seen as tools of conservation and ecological awareness. "What we try to do is to help our guests see how this animal fits into an ecosystem," says Ramirez. "The actions we have right here on this lake have to do with this animal's natural habitat thousands of miles away. If you care about an animal you have to do more than care about that one individual, you have to learn to be concerned about the whole ecology: the fish that animal eats, the shoreline, the pollution in the water. It's not hard to get people to care about a killer whale, a panda, or a condor. But habitats are being destroyed—people are less motivated by the environment as a surrounding whole. That's why we have recreated the geology and botany of the Arctic right here, indoors."

Ramirez doesn't mind the public critique of aquariums. "One of the things that the protest movement did is put a lot of scrutiny on our facilities. Today the type of people who work here are very different than they once were. These are very compassionate, caring individuals. There is no one who cares more about the health, happiness of these animals than those of us who work with them on a daily basis. We and the animal liberation activists care about the same things, but we diverge somewhat on how to reach our goals. Maybe we should all work together and quit fighting each other!" He has since been dismissed from his position at the Shedd.

I assemble all my gear and lay it out on the wet concrete floor. The whales are already giving me curious looks. The hydrophone is dropped down inside a mesh net so they won't try to swallow it—I'd heard some other aquaria had lost some equipment that way! The underwater speaker is only a few pounds in weight, but we're advised to hang it off a rope rather than let its own cable hold it down there. The rest is above water. I put on a pair of full-coverage padded headphones and immediately hear a rumbling rush of scratches and waves. Who would imagine a calm aquarium could produce so much noise? It's an intimate setting, a far cry from shivering on the deck of the sailboat on the Johnstone Strait.

I spend three long mornings wearing these phones, listening in on the underpool world. There are sporadic beluga honks, whistles, burps, and shrieks. I try the clarinet, and the louder, more glistening soprano saxophone. Sometimes the whales seem to notice that shinier instrument. It's louder, and a bit more impressive to look at. They come out of the water to check it out up close. I don't play it as well as I do the clarinet, so for me trying it is more of an homage to Paul Winter and Wayne Shorter than anything else.

On the first day, no response from the whales. But Ken thinks otherwise: "Look, every time you stop, here comes our pregnant whale. She always looks at you while you play, and she always comes over when you stop."

I wanted more. "Is this as noisy as they get?"

Ken smiled, "Oh no, sometimes it's just a cacophony of sound here. So loud you can't hear yourself talk."

Second day, many hours, quizzical whale looks, hours of recording, still no audible answer. I just keep trying.

Only on the third day do I get a result. One particular note, a G just above middle C. The pregnant whale hears me play that sound, and a few seconds later, she appears to copy it almost exactly. Look at this sonogram in figure 10, and you can see the similarities and differences between the sax note and the whale note. The clarinet is in gray and the whale is in black:

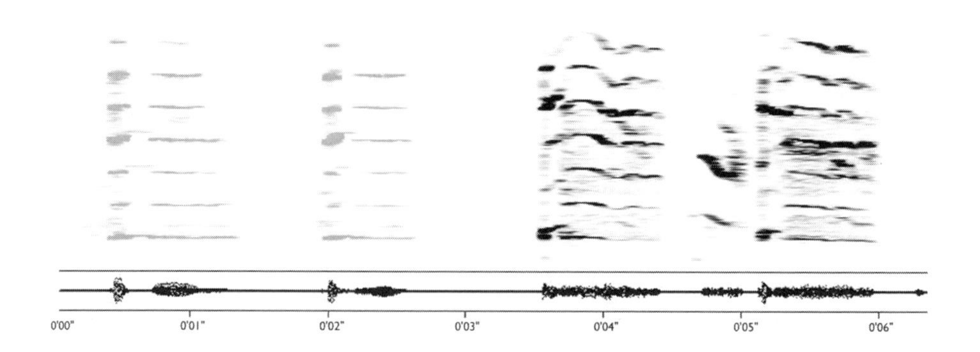


*Fig. 10. The clarinet plays two notes in gray; the beluga copies them immediately.*

I play "*buweeah, buweeah*," and I hear back "*heyaaah, heyaaah*." The sonogram shows that the overtone structure, the timbre or color of the whale's sound, is quite close to what I am doing—not just the pitch, but the phrasing as well. To me the whale has definitely listened and learned. Why? Does something about that sound matter to her in particular?

Belugas are well known to play games with sound. "I worked at another facility once where we used a dog whistle to let the animals know that we liked what they did," says Ramirez. These whistles can be adjusted so they are within the range of human hearing, or so high that only dolphins and whales can hear them. "The belugas would watch our dolphins in our performances, and they would imitate our whistle so they could manipulate what the dolphins were doing. They seemed to time it well so it would cause the dolphins to mess up. . . . Perhaps they had the pleasure of watching a dolphin get confused!"

Maybe they hoped to confuse me too. I travel through music, and I would travel with these cetaceans and see where the sounds would take me. On that third day in Chicago I got six seconds of hope. Later the clarinet and beluga seemed to interact around a coherent rhythm:

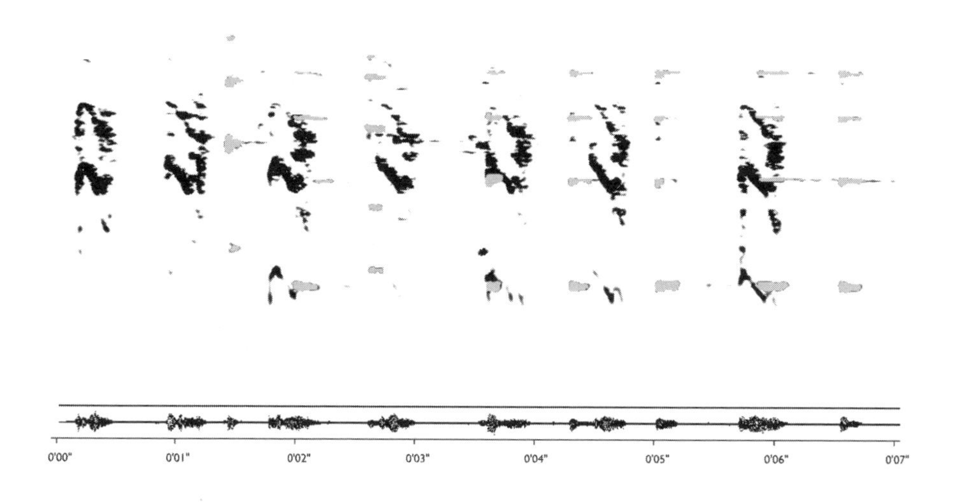


*Fig. 11. Rhythmic interplay between clarinet in gray and beluga in black.*

The gray, more pure tones in figure 11 are the clarinet; the resonant whistle-honks in black are the whale. I see a common pulse here, though not as much as in the first example. Here instead is a more complementary, rhythmic engagement.

Belugas make more varied sounds than any other whale, many easy for humans to hear above the water. Because of the vast complexity of beluga sounds, it has proven difficult for humans to categorize them well. Cheri

Ann Recchia described the basic kinds of sounds made by the captive belugas she studied in an aquarium as clicks, yelps, chirps, whistles, trills, screams, and buzzsaw. In the early 1980s, Canadian scientists Becky Sjare and Thomas Smith conducted an exhaustive survey of the vocal repertoire of wild white whales summering in Cunningham Inlet off Baffin Island, and they found clicks, pulses, noises, trills, and a variety of whistles they described in great detail. Some sounded like fragments of scales, with clear pitches, and others were whoops and cries, rising up, falling down, in various clear patterns. Some warbled all over the place. They found no real correlation between particular sounds and particular behaviors.

The whistles, which may have signature qualities for dolphins, seem in belugas to be closer to the forms and shapes used by killer whales, equally easy for humans to hear. Whether they have the same clan-identifying functions is unknown. In fact, for all the statistics collected around beluga sounds, their complex music or language seems more inscrutable than the code of any other whale.

My brief interactions with the whales have put me in a good mood. (Perhaps I'm easy to please. You could get that much with birds just whistling out your kitchen window.) Mostly it's the strangeness of the experience of wearing headphones for many hours and listening to the noisy rumblings of the underwater world, straining to hear the sounds of sea canaries bounding from pool to pool, bouncing off the blue-green walls, subjects in benign experiments where they are endlessly loved by passersby. True, one might also feel sadness at the sight of cetacean prisoners in a castle moat, with gawkers harassing the trapped white whales. But I'm more with Ken, feeling playfulness in the beasts, a tendency to explore. *Hreeaaah! Hreeaaah!* What would their cousins do in the wild? It was time to go to Russia to find out.

Rauno's plans are coming together. "I don't think the white suits will yet be ready this summer," he apologizes. "We are going to focus on *your* interests, playing sound to the whales. I am really curious as to what they will do."

Officially, Rauno is a physicist, designing experiments for the new supercollider at CERN in Geneva. But he has been interested in whales ever since he took his eight-year-old son whale watching in Norway fifteen years ago. On the drive home to Finland the boy said, "Papa, couldn't we do something to save those whales?" Rauno thought a moment and decided, "Yes, we can. It won't even be all that difficult." This was in 1990, the very earliest days of the internet, at first a way for scientists to keep close contact with each other all over the world. The Web itself was developed at CERN, and one of the very first not-quite-academic web pages was Rauno's "whale watching web," which he still maintains (www.whaleweb.org). Even now the site looks like one of the world's first web pages—all text, no animations, pop-ups, or ads.

"To me the internet is just a map," says Rauno. "At first my site was just a place to present information on whale-watching companies all over the world. Then it expanded with pages on whales in literature, proverbs, pictures, and sounds. If I look back to my own childhood, I remember a photograph in a Finnish *Readers Digest* of the dolphin brain compared to a human brain, and they were so similar. An article by John Lilly, I believe."

When not ensconced deep in the laboratory, Lauhakangas travels around the world, often to areas frequented by whales where the local human inhabitants haven't given much thought to the great animals as assets to tourism. Rauno then helps the locals start up whale watching as a locally owned industry. He's done this in the Azores and Iceland, and now he hopes to do the same in Karelia, one of the lesser-known Russian republics. Just east of Finland, it is well within reach of thousands of potential ecotourists.

"We are taking along a photographer and videographer," he announces. "We will document the whole experiment and spread the word that the white whales are out there. A few groups a season would be enough to bring some prosperity to the tiny village of Kolezhma, from which we will set out onto the White Sea."

"Only a few tourists would be enough?"

"The average annual income in this place is probably a few hundred dollars a year. This will be the beginnings of capitalism for this beautiful place. And the market economy is the most powerful vehicle for change that humans have yet invented."

"This time I have got the documents certifying that none of our equipment is radioactive," Rauno smiles, as we are scrutinized by Russian border guards near Joensuu. "I forgot those last time." The journey to

Kolezhma from the Finnish border takes about two days driving on roads of questionable quality. We pass through the capital city of Petrozavodsk, where no one has yet bothered to take down the statue of Lenin, and there is still an avenue called Pravda Street. In a late-night student bar a guy leans over to me and says, "You know, it is so nice to hear English spoken here. No one ever comes to visit this city, and it is quite a nice place, don't you think? Seven new outdoor cafés opened up on Lenin Avenue just this year!" It has been a long time since I've been anywhere outside the United States and met someone actually excited to see an American.

North the next day the drive is nearly all through deserted spruce forest, home to little more than undernourished moose and squirrels. We turn along the Stalin Canal, dug by hand in the early days of the Russian Revolution, a massive testament to forced human labor. It is a man-made river of tears. "We prefer not to dwell on this history," says our young guide Alexander Velikoselsky, who is practicing his juice harp to play to the whales. *Boing, boing, boing.* We pass a crumbling museum of petroglyphs, a concrete edifice built to protect an important story-rock, now boarded up because there is no money to keep it open. Russia has many more things to worry about than its history, whether ancient or modern.

But the door is ajar, we wander in. No windows, only bars and cracked glass. The glyphs themselves are protected under sawdust and burlap. "Never mind that," smiles Rauno. "The most interesting petroglyph is not on this rock, but in the woods just one kilometer from here. I saw it in a book years ago, but found the location by chance. I brought Jim Nollman here six years ago and we made this rubbing":

"The location is a carefully guarded secret. This image on the rock is our holy grail."

"Why do you say that?"

"Well, what do you see here?"

"Well I can see the nose, that's definitely a beluga, perhaps breaching the surface of the sea. A hunter has his arms and legs flayed, he's ready to strike with his spear."

"Look again," Rauno grins. "In science we look for the root of things, the truth extracted out of all the extra information."

"All right, let's just say it's not a spear. Maybe it's a didgeridoo, he's blowing deep buzzing tones into the water. But the natives here never played any instrument like that."

*Fig. 12. The mysterious petroglyph linking human and beluga.*

"Well," counters Lauhakangas, "nearly a thousand years ago we have the testament of Bishop Adam of Bremen, who was the first literate European to chronicle these regions. And I quote: 'All people in the Far North are Christians, *except* those who live on the coasts of the Ocean above the Polar Circle. It is told that they have the wisdom to know what each one is doing in whatever part of the world; and they call whales to the beaches by murmuring powerful charms. They know many things by experience, which are only told in the Bible about wizards.' So long ago, the people by this Arctic sea may have spoken with the whales. No one alive remembers."

Gazing across the road to the old canal I can see why. The recent history is so much darker, built on so much death. In the twenty months it took to build this canal, two hundred thousand people died. The petroglyph seems to come from another world.

"Notice what is in his hands!" glares Rauno. "The rattles of a shaman. He looks much more like he is dancing than hunting."

"I still think the natives would have had a much more prosaic use for that whale. They wanted to eat him. Shamanic or not, animals first and foremost meant food."

"So whose fins are between the legs?" Rauno shot back.

"What legs?"

"I think you're missing it completely."

"How do you mean."

"Notice her breasts."

"Whose breasts?"

"Don't play dumb, that is a woman giving birth to a whale!"

The Chukchi of Siberia speak of a woman who fell in love with a bowhead whale. This bowhead saw her walking along the rocky shore and turned himself into a young man. He would stay for a while, then return to the sea, disappear for a time, and always come back. This species, like the humpback, sings a plaintive song, one phrase up, one down. *Whooop, eroop, whoop, eroop.* The woman was entranced by this melody and could not forget it.

The woman who married a whale gave birth to human children and whale children. The boys and girls played on the rocky beach in the sun. The baby whales swam in the lagoon by the village, but when they grew too big, they would disappear out to sea and join the pods that swam by the village a few times a year.

She would always tell her human children, "The sea gives us our food, but remember your brothers the whales and your cousins the porpoises live there. Never hunt them, but watch over them. Sing to them."

Her children grew up, then they had children of their own, all human. The village prospered until one very tough winter. There was little to eat. One grandson told another, "Why don't we kill a whale? There's certainly enough meat and fat on even one to get us through this season."

"Remember what Grandma said," replied his human brother. "Those whales are part of our family. We must leave them alone."

"What kind of brothers are they?" said the other. "They are long and huge, they live under the sea, and they don't know a word of human speech."

"But we can sing to them, and they listen."

"You sing. I'm not going to die of starvation." With that he paddled out into the sea. Soon one whale swam slowly up to his boat, as they were used to doing. It wasn't very hard to spear him.

When they dragged the dead bowhead back to shore the killer went to his grandmother, proud he had found food to save his people.

"I killed a whale, grandmother. There is meat and blubber for all to eat." The woman who married a whale already knew what had happened. Then she cried. "You killed your brother, just because he doesn't look like you."

She closed her eyes and died.

The Chukchi sigh. It all went downhill from there. Now even when a human kills another human, no one is really surprised.

It could have turned out differently. Maybe the woman who married a whale gives birth to a child with a special sensitivity to the sea. Maybe he grows up to be like the ancient Greek musician Arion, who played his lyre to the dolphins of the Mediterranean, belugas of the warm waters. They were always happy to see him and swirled around him whenever his music began.

One day Arion was captured by a band of pirates, who granted the poet one last wish before they made him walk the plank. He started to sing the dolphins' favorite song, and as he was pushed into the water they gathered around him, ready to carry him safely to shore. The pirates had never seen anything like this, and they were moved to give up their wicked ways, spending the rest of their days singing and dancing, having lost the will to kill and fight.

That was six thousand years ago, just two thousand years before this petroglyph was carved into the Karelian stone. Edmund Spenser encapsulated the story in a poem:

> Then was there heard a most celestial sound
> Of dainty music which did next ensue,
> And, on the floating waters as enthroned,
> Arion with his harp unto him drew
> The ears and hearts of all that goodly crew;
> Even when as yet the dolphin which him bore
> Through the Aegean Seas from pirates' view,
> Stood still, by him astonished at his love,
> And all the raging seas for joy forgot to roar.

Albrecht Dürer drew this same moment, but his cetacean is a kind of vicious-looking evil beast, more like a sea warthog than anything real. Philip Hoare recently wrote a whole book inspired by this image entitled *Albert and the Whale*:

*Fig. 13. What Albrecht Dürer thought a dolphin looked like.*

It's a far cry from the gentle inquisitive faces of the smooth, white belugas who we were hoping to attract with sound, and a long way from Rauno's petroglyph. (In 2021 the great whale chronicler Philip Hoare wrote a whole book inspired by this image called *Albert and the Whale*.)

Maybe the woman *is* giving birth to a whale. Maybe a shaman is becoming a whale himself through magical incantations, murmuring charms. "Of course," admits Rauno, "the early-twentieth-century treatises on Karelian petroglyphs just describe this picture as a hunter pursuing a whale. They had a more limited imagination back then. I believe each generation interprets ancient art in a new way, to meet its needs. We need the whales for something else today."

At the edge of the sea, in the crumbling industrial town of Belomorsk, we turn east on a tiny dirt road for the two-hour drive to Kolezhma. The farther we go, the emptier the landscape becomes. Nothing but close-knit evergreen trees, mile after mile. The dirt track crosses streams and swamps, on rickety bridges made entirely of wood, huge straight timbers. We stop to check each one for missing slats before we proceed. They all seem okay, some just barely so.

At the end of the fifty-mile road is the outpost of Kolezhma, more beautiful and distant than one could imagine. Everything is made out of weathered, unpainted wood. There is one tiny shop that is closed. It's not clear what people can do out here except endure, keep going, grow their own food, and stay alive. Several hundred remain here, and hardly anyone leaves.

"Look at this, we're back in the Russia of nineteenth-century novels." Rauno has big ideas. "This place is going to be the whale-watching capital of the White Sea!" he gestures wildly. "There's nowhere in Finland like this, people will pay many euros to see it, and that's the only choice," now with a sinister grin, "because nothing can stop the green snake of *Das Kapitalismus* from rearing its ugly head."

Green snake?

"The whales as they are can bring prosperity here. Come, I will show you! The boat is ready for our transport to the island of Myagostrov. It's a pagan holy place. Show some respect, Anna," he gestures to our photographer. "No woman has ever been allowed on this island before."

We load our gear into a rusty, green metal boat that resembles an above-water submarine, hurrying aboard because the trip can only be made when the tide is still high. The water's surface is as smooth and gray as the sky. It looks like it should be cold out but the temperature is nearly ninety degrees. The Arctic gets warmer than you think.

It's a slow two-hour boat ride to the northern tip of Myagostrov Island. Sitting inside the hull I feel I'm stuck in a World War II submarine. The rocky point is marked "Cape Beluga" on the map, since this is one of the seven known spots in the White Sea where belugas congregate in summer. A mile to the west we find a series of weather-beaten cabins; one of them looks brand new.

"That's the sauna," beams Rauno. "They built it all in one day last week, just for us."

"Jesus, I hope it's not *too* Russian," says Gari, our video man.

"What do you mean?"

"You know, the kind of place where they also keep animals or hang the salamis up to dry."

The boat can't get too close to shore, so we anchor a hundred yards out, take off our shoes, and start carrying in the huge amount of gear we've got box by box: boat batteries, invertors, video equipment, five cases of beer (one for each day, essential for all traveling Finns), sleeping bags, food, hydrophones, saxophones. We're walking through mud and over slick, smelly, algae-ridden rocks. The shore is awash with mosquitoes and flies. "Careful," warns one of the Lechki brothers, the guys from the village who have organized this trip for us, "the woods are full of tiny poisonous snakes."

Inside the cabin it's even warmer than outside! Who knew you could sweat this much so close to the Arctic Circle? At night it's so hot and the bugs buzz so ferociously around our ears that sleep is well-nigh impossible. Plus it never gets dark. I'm hiding in my down bag to keep the bugs at bay, perspiring profusely. Who could think of a sauna in this weather? I keep telling myself as soon as the hour gets reasonable, I'll just get up and plunge into the sea. Finally it's 4:30am—that seems late enough. I jump out of the bag, and dash outside, the sweat dripping off me as the mosquitoes scatter. I can run, but cannot hide. The sea is a quarter mile off, the tide is so low. Nothing to do but wait.

As the sun rises high we're too excited to complain. We take a dinghy over to the cape to install the hydrophones and the speaker. It's a different world over there, smooth pink granite, a nice breeze, just a few deer flies to reckon with. The hydrophones are dangled from buoys with ropes. The underwater speaker is suspended from a pole that looks like a broken fishing rod. The tide comes quickly up.

On the hill above us we notice a small wooden lean-to. Three figures stand there, silently watching us. A ten-minute run up the lichen-covered boulders and we find Russian photographer and whale expert Alexander Agafonov, a bushy-bearded fellow with a wide-brimmed hat and a plaid shirt. We expected he would be there, along with two young assistants. He was sent by the Shirshov Institute of Oceanology in Moscow to spend a month watching these very same belugas from this craggy hilltop. I offer him a copy of the book *Dolphin Societies,* which contains several papers from the Shirshov Institute translated into English on the behavior of dolphins. Agafonov is one of the authors, and he's never seen this volume before.

"*Da*," he smiles. "We will watch and listen to the whales. And we shall listen to you making music with the whales. It will be interesting to all."

On the horizon to the north I glimpse a low trace of land, about seventy miles away. The Solovetsky Islands were the central administrative facility of the Soviet prison camp system. Solzhenitsyn called them the Gulag Archipelago. Now the offices have been returned to their original purpose, a key monastery in the Russian Orthodox Church. It is another good place to watch for beluga whales.

I get out my soprano saxophone, since that worked best at the Shedd. Put on the headphones, listen to the rumble and *thlack* of wave against rock. This is no placid sea today. I take in the noise, wait, and wait. Then, I hear them. Rasps and squeaks, clicks like a strange radio broadcast from a planet far away. "Guys, get the cameras out, they're coming."

The whiteness of belugas is beautiful, but anyone steeped in whale lore will remember how Ishmael tells us the very color of Moby Dick the White Whale is his most terrible feature, "not so much a color as the visible absence of color, and at the same time the concrete of all colors . . . a dumb blankness, full of meaning—a colorless, all-color of atheism from which we shrink." The White Sea is not white at all but a dull gray, or a deep colorlessness, that adds in hollowness with thoughts of the terrible human history played out on these shores: war, incarceration, torture, fear. That's all over, and we're lucky the white whales remain to remind us that nature can be pure. Whales do not do such terrible things to each other; that's why John Lilly thought they were far more intelligent than we are.

The younger whales are gray, blending in with the sad surface of the sea. Through the headphones I eavesdrop on their noisy, vibrant world. Above there is the heavy thumping of the waves bashing on the rocks, in splendid downward sweeps. High whistles at the limits of human hearing. Pings and bleeps, new senses of rhythm and order, new beauties in tone and kick. I strain to pick out the whales above the wash of underwater noise. White noise in the depths of white noise. Rauno says it's a bit like trying to find new particles in the printouts from linear accelerators. "Although in physics," he smiles, "there is a lot more noise than this. A hundred times more."

Is there anything musical in these sounds? Easy to say no: they're clicking to make sense of the things they encounter in a dark environment. Belugas can hardly see, but they can clearly detect the outlines of an object behind an underwater wall. The military knows these whales have amazing detection abilities. Their echolocation, far more powerful and directional than dolphins, enables belugas to find their way through cloudy seas under noisy, breaking ice. They click to find their way, and whistle to signal; signature vowels and pod. What's musical about that? Foreign languages are inscrutable, able to be decoded. But foreign music? It still has a beat we can follow.

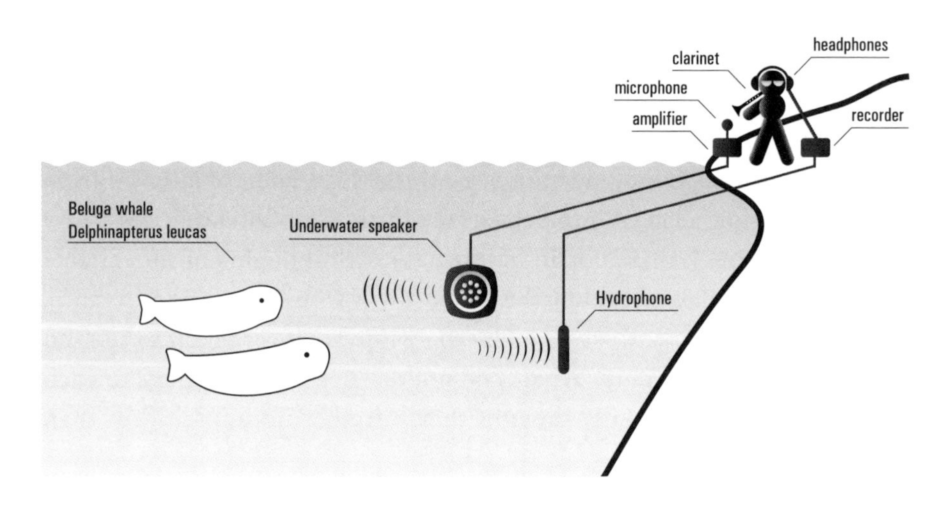


*Fig. 14. How I played along with belugas while standing on a Karelian shore.*

A whale sings, a clarinet rings. The sounds overlap and connect. I smile, listen, and play again. I imagine the whale is responding to me, yet another human conceit. Why would a white whale care for us? They hardly know we're here, near the Arctic Circle, where a steady breeze offers some relief from the sweltering heat. I can't sit still, I have to get up, move around, dance a bit while I reach out to the whales, remembering an old Karelian proverb: *Kundele korvilla ela peržiel*, "Listen with your ears, not your ass."

Agafonov comes running down from the hilltop. "Yes, we have been listening. The belugas are definitely responding. Perhaps music is a better way for humans to communicate to these whales." The tapes will be sent to Moscow. The great Vsevolod Bel'kovich, doyen of Russian whale science, will listen to them, and analyze. We will find out what's going on, we will get to the bottom of this. Whale-human music has begun.

The belugas scream, wail, cry, and click. It is an alien improvisation, within a strange style that is right here, reachable in our time. The more I listen to the belugas and find my own multiphonic shrieks that merge with theirs, the more I learn to taste this new underwater music. John Cage called the interconnected patterns of sound all around the strands of a vast natural symphony, where overlapping intention forms the music of what happens. Days alone with headphones trying to reach out to belugas and my whole notion of what can be music begins to change. I listen beyond the edge of my species, trying to find my part in the underwater soundscape.

We spend several days in the sweltering cabin and eagerly wake up every morning to walk along the shore to Cape Beluga. When the sea is rocky and rough, the waves sharp and white, whales do not appear. When the water is smooth, the distant islands seem to float on the surface. At high tide, the whales appear, right on schedule. It might be night or day, but it is nearly always light this far north. When they hear me play, some kind of aural response happens. This isn't science, so I can't be rigorous or conclusive about it; I have hardly any data. I listen to decide if the result of our encounter is musically interesting.

At one point I try the same middle G the whale in Chicago copied after hearing it for three days. Right away there is a response! Either that sound is easy for belugas to master, or it is already a tone that means something to them. Either way, I feel as if I'm getting through. The whale and I share a sound for a moment or two.

The project is getting results. "This I like," beams Rauno, lying back on the warm rocks enjoying the sun. "I need this kind of experiment as a break from all that tinkering in the lab."

"You speak often of *Das Kapitalismus*, Rauno, but you don't seem much interested in making money yourself."

*Fig. 15. Rauno holds the microphone with his toes while I play a* furulya.

"Capitalism is like gravity. Somebody starts to figure out how to make money, his brain will be taken over by *Das Kapital*, and there is no time to think of anything else. I do not like that."

"So capitalism makes things happen if we don't let it consume us?" I guess that is what Marx was actually trying to teach us. That idea is already in *The Communist Manifesto*.

"People cannot be fenced in this free world, but they can be educated." Rauno stands up and stretches, then starts to pack up his stuff. "Now we have to get back to the village, the Kolezhma Beluga Festival is upon us."

"What festival?"

"Oh, didn't I tell you. We're going to have a little event for the village, show them what we've been up to. When I mentioned this to Elena at the Karelia Tourist office in downtown Petrozavodsk, she was at first concerned, 'What, a festival in Kolezhma, population three hundred? Well, why not! Let's make it a tradition.' You and I must invite Agafonov as well, to give an impromptu presentation. Look, I've made a poster for the occasion":

"See, you can zoom in on Google Earth right onto the village. Look at those long buildings from above in the wilderness. You think they are secret military installations? I heard they are old mink barns . . . They used to feed belugas to the minks to fatten them up. I left the time blank so I could surprise you."

Rauno is never at a loss for surprises.

When the boat comes to pick us up we hear there has been a terrible storm in Kolezhma. Power has been out, several roofs have been ripped off their frames. The Lechkis look quite shaken. Nature here is more extreme than it looks. On the journey back we are all a bit apprehensive, but when we arrive at the village the sun is out, everyone seems in a good mood. The store is open today, and we stock up on cakes and vodka. As we carry our equipment to the village hall, at every turn more little boys turn and follow us. By the time we reach the old wooden door we have quite an entourage, like pied pipers of whale music.

As I take out my equipment on stage, the little boys refuse to leave. They want to be part of the show, sitting beside me, grabbing at the flutes and the microphones. All of a sudden the room goes dark—the power is out again. A generator is found. Power returns, but with a motor's rumble. I play the sounds of belugas and other whales, making music with and around them. Sasha Velikoselsky joins on juice harp and Gari Saarimaki is on guitar. More than a hundred villagers are here, nearly half the population. Not much else going on in town that night I guess. The smiling whales glisten and sing.

Agafonov is next, and he speaks softly and carefully on why he has spent a month doing nothing but watching these white animals. He shows a video in which two groups of four parallel whales approach each other, turn away from the shore, and then all go off together in a line of eight. It looks very organized. His team has collected careful data on the movement and appearance of the whales, every day. Rhythmic, repetitive work. Great patience is needed to conduct it.

Then Rauno introduces the Russian language version of a film he helped to make with Canadian director Patricia Sims, *Beluga Speaking Across Time*. This documentary begins with our petroglyph and goes on to describe differences in the way Canadians and Russians understand the beluga whale. In Canada the natives still hunt and eat the beluga; in

Karelia that is long forgotten. The film shows Bel'kovich in his leather jacket getting out of a black Volga limo, going into his laboratory to reveal the various James Bond–type contraptions he has invented to study dolphins and belugas. Those were the days.

After the festival is over, the village returns to quiet and we retreat to the Kolezhma Cultural Center to unwind. Vodka is passed all around. Rauno proposes a toast: "To the beautiful icon from Solovetsky Island, depicting the founding of the monastery, a pilgrimage site for all devout Russian Orthodox, by Gherman and Savvatiy of Kirillo-Belozersky, arriving across the sea accompanied by two white beluga whales."

"Wait a moment," says Svetlana, one of Agafonov's assistants. "I worked for six years in the archives of the Solovetsky Monastery. There is no such icon painting, there is no such story."

"*Kippis*," smiles Rauno, holding his glass up again. I know he has a way with the truth. Each generation reinterprets history for its own ends. And in Russia no *salud* is innocent.

"To your shamanic sounds crying out to the belugas," intones Agafonov. "Like the throat-songs of Tuva, the quality of the sound matters more than any meaning."

After another shot of vodka I glance out over the thick forests to the colorless sea, with another Karelian proverb in mind. *Mägi mägenke ei yhty yhtei, a ristsikanz ristsikanzanke yhtyu*—"Mountains never meet, but men do."

We may all be friends now but there used to be many walls between the East and West, and these walls extended to the world of dolphins and whales. If the Americans were tapping dolphin intelligence for nefarious military purposes, then you can be sure the Russians were too. And they knew of the echolocation abilities of belugas for a long time, so the Russians were training their secret underwater weapons for decades before any Americans knew these whales had vowel sounds. As we had dolphin facilities in Mugu and Loma, they had their animals in the Black Sea off Crimea. We only know about this because one got out.

In 1992 vacationers on the Turkish coast of the Black Sea were treated to an astonishing site: a beluga whale, one that seemed unafraid of humans and eager to take fish from people's hands. Such an animal was

# Kolezhma Белужа - festivaalit 12.7.2006 klo. __.__ Doma: Kolezhma School

David Rothenberg, clarinets, saxophones
Gari Saarimäki, guitar
Aleksander Velikoselsky, guitar
Myagostrov beluжa, seireenilauluja
____________________, Pomorilaulut

Elokuva: Белужа: Сквозь призму времени 50'

*Fig. 16. The Kolezhma Beluga Festival*

totally unknown in the wild in that part of the world. Everyone assumed he had escaped from an aquarium, and the only one somewhat close was a former Soviet military facility in the new Republic of Ukraine, a thousand miles away. Was this whale a Communist spy? What military secrets did he know?

The first white whale to appear in Turkey became a national sensation. Schoolchildren all across Europe and the Middle East learned his name, which the media had decided was now Aydin. Soon an official message came from the government of Ukraine: "Yes, that whale escaped, and he belongs to us. His name is Briz, he was born in the Sea of Okhotsk off Sakhalin in 1984, captured by helicopter in 1987. We are coming to Turkey to get him."

As in the case of Keiko, the children of the country mobilized in support of the whale. Could money be raised to support his rehabilitation and return to the wild? Nobody wanted him back in a military facility in the Crimea. Except for the Russians, who sailed their ship during the night to Turkey and collected the whale to bring him home. But more and more children from all over Europe wrote letters to the Ukrainian government. Look, said the Ukrainians, we've put that whale in an oceanarium, where he is performing shows for little children like you. He's not in the Navy anymore. And if you want him, just give us $80,000, the going rate for a beluga in Japan. Although this was a tiny fraction of Keiko's price, none of this whale's fans had enough money.

As luck would have it, a huge storm lashed against the Crimean coast, and Aydin/Briz escaped again. The whole saga was beginning to sound like a Lou Reed song. The Soviet beluga swam right back to Turkey, and the nation was overjoyed. His picture was in all the papers. Restaurants sprouted up called White Whale, and this national hero was praised as an alien who truly deserved Turkish citizenship. Later that year the Turkish government passed a law decreeing that no white whale can ever be captured again in Turkish waters for any purpose, and to this day at least one beluga whale freely roams the Black Sea.

Moscow in winter is famous for bitter cold and heavy snow, but this winter, as in much of the world these days, it is warm enough to be only gray and rainy. I see none of the opulent excess too often described in the Western press, just crowds of people shuffling through gray streets, crumbling buildings, heavy traffic, and the sense that here is a system that no longer works.

In December 2006 I went to Moscow to visit my whale scientists friends. The subways are grand and precise, Stalin's pride. They still run on time. The rush of bodies underground is like nowhere else I have been except New York. I've come to visit Agafonov, who, when not watching Arctic whales, lives with his mother in a tiny apartment. "I'm fifty years old, and I've been married twice," he smiles. "Neither of my wives could appreciate how much work I have to do." Agafonov studies belugas only part of the time. His main job is at the Russian Union of Art Photographers, of which he is the president. Since the collapse of the Soviet Union, there is no longer much support for culture or science. "But we are free, and this is much better. We Russians always find a way to get by."

The photographers' union used to be in a grand old house, but now it is in the basement of an apartment building near the Tekhnika subway stop. Nicer, says Agafonov, than the Oceanology Institute. Here he has gathered several of Russia's best whale researchers to present their work and hear what we found the previous summer in the White Sea.

We shuffle into a dingy apartment building and head downstairs to a basement office crowded with coffee-table photo books and posters of famous images of Russian history. "Under Communism our office was in a beautiful mansion," Agafonov admits. "Now we must be content with this." On a desk are bowls of pickled herring and plates of salami. Bottles of wine and vodka await us.

Vladimir Baranov shows a video he made off of Solovetsky, with a special seafloor-mounted camera, of svelte white whales nuzzling each other in the White Sea. Beluga mating in action. "Da, look he slidez up next to her and then sticks it in, voila! Then he svims away, and back, does it von more time just to be sure!" The distorted grainy color looks like something from an early cinema archive, even though it was just made a few years ago. "Zhis video is especially popular with the ladies," and he takes another swig of vodka.

Roman Belikov is one of the brightest young whale scientists in Moscow, back then just finishing his Ph.D. He presents the comprehensive work on beluga sounds he has been doing along with Bel'kovich, the most extensive analysis of these sounds in any country. Bel'kovich himself is in the hospital and cannot join us. They have begun to draw some conclusions. Bel'kovich and Belikov found six basic types of beluga sound: "creak, bleat, chirp, squeak, whistle, and vowel." In social interaction between whales there is more bleating, chirping, and whistling. There is plenty of chirping during beluga sex, but absolutely no bleating. The most vowel sounds are heard during peaceful swimming.

The vowels are the one kind of sound the Russians found that was not similarly categorized by Canadian scientists. It is a sound few other researchers have identified in any other whale, or any other animal, for that matter. It's an ingenious way to give a complex, noisy sound a clearly recognizable identity, like the vowels of human languages. The particular shape of the overtones in the sound, called "formants" by speech pathologists, are what give each vowel its quality. In the belugas, the Russians heard these basic vowels: *ah*, *eh*, *uh*, *o*, *u*, and *ee*. That whale in Chicago who copied my saxophone note seemed to be making an *eh* sound. Bel'kovich and Belikov believe these vowel sounds can be the key to identifying individual whales interacting in a group, because they seem to be distinctive, not shared. Instead of the signature whistles found in dolphins, belugas may have signature vowels.

I play my two best recordings from the previous summer. Unlike the orca duets, which were simple and direct, these beluga encounters are windows into a much more cacophonous underwater world. On the first recording, the belugas whistle and growl in the midst of the noisy *thlacks* of waves against rock. The clarinet sounds like an interloper in a world of crazy noise. No sound they make can be characterized, no tone I try out has a place. We're playing at and around each other, and I am slowly lured in.

In the second recording the human world is muffled. There is wind and a tentative clarinet. Once again there is that searching note G. Faintly in the background, in the other channel, the whale answers. I've carried my one syllable from Chicago to Myagostrov, and I hear it echoing back. Months later now in Moscow, the same note resounds in my head. It's there, but it's hollow. We don't know what it means.

I ask the Russians whether music has any place in the attempt to understand the world of beluga sound, since it seems that Western scientists have cast such a woolly approach off decades ago. "We believe," says Belikov, "that some of the sounds of belugas are specifically indicative of their emotional state." Does a whale with such a wide range of possible sounds have more emotions than those with a more limited repertoire? Not necessarily, but they certainly have a greater range of expression. If we decode such sounds, will they lose their allure and turn into message in our ears? Of course not. Vocalizing is what belugas do, it is their very nature.

The beluga, like Moscow, does not believe in tears. But they do believe in sound. It takes a little more effort to find music in it, but when we do, beauty starts to appear. "Beluga sounds are a very complicated subject for investigation," says Belikov. "Perhaps we will one day be able to know as much about their meaning as we do about orca sounds, but I have some doubts. If everything would be so easy than all problems would have been already resolved. Everything is very elusive in beluga signals."

Agafonov presents the sounds and videos he has assembled from the previous summer's extensive observations. It is a catalog of fabulous noises that no one understands. The clarinet and the whale? "I'm afraid there is not yet enough data. You must come back with us next summer. And maybe the summer after that."

Baranov pours another round. "Let us watch the film again, this time with music only, very relaxing," and a strange dissonant soundtrack now comes on, with dark electronic chords. The whales cavort around the camera, look us in the eye, and wonder what strange machines we will come up with next. A middle G swirls around in my head; the whales cry up and around it, occasionally copying that one simple tone.

What a pleasure it was to take this trip to Russia in 2006. I am so sorry that because of calamitous world events, it would be impossible in 2022. We can only hope such journeys will be possible again soon.

Rauno offers up his latest discovery, an ancient song from the Finnish folklorist Elias Lönnrot, compiler of the national epic *Kalevala*. Deep in the midst of his collected writings, the whale-watching physicist has found a whale lyric, the single reference to them in hundreds of pages, lying in wait for someone to notice:

There are three high mountains
Three high falls
Three eddies in the water.
A fire on each eddy
A man beside each fire
 A piece of whale in each man's body.

"And that doesn't mean we should *eat* them," Rauno raises his glass once more. "In the White Sea they remember us, and their songs have found a place in our hearts."

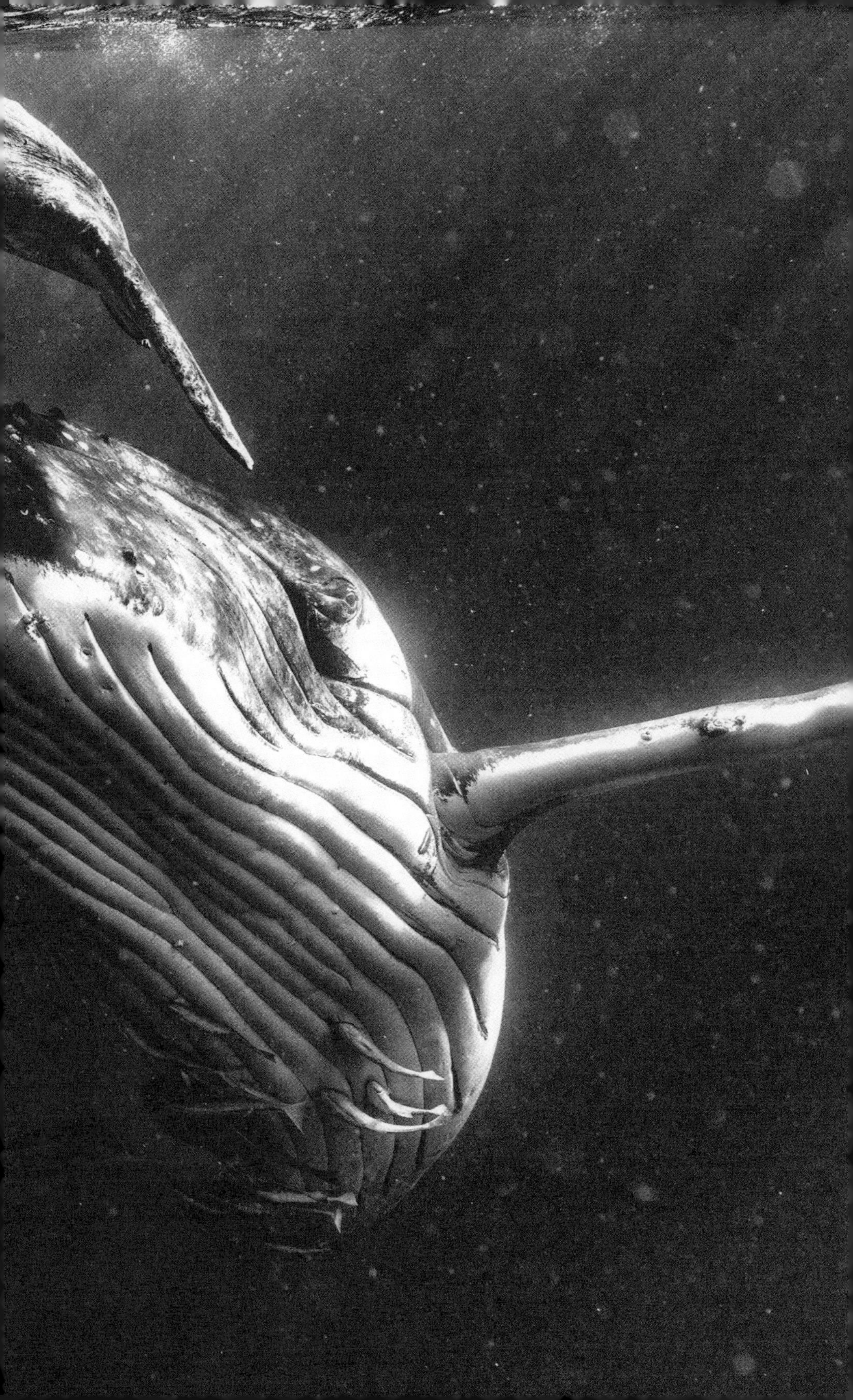

# 6

# THE LONGEST LIQUID SONG

## Humpback Music and Its Changes

Evolution does not just encourage the survival of the fittest. It produces wondrous beauty and strange ways for animals to be in touch. Cuttlefish change their shape and color to engage with each other. Penguins identify each other through slight variations in their calls. Elephants tap their feet to each other from miles away, their vibrations carrying for long distances under the ground.

Communicating with symbols is rare among animals. The most famous example is from the honeybee, who represents complex information with its famous waggle dance, indicating the location of blossoms where pollen can be found. Bowerbirds in New Guinea fashion sculptures out of twigs and leaves to attract the attentions of potential mates. The males of some species even spend hours a day painting their works with crushed berries, or decorating them with flowers of a particular color. These creations are not nests, but artworks devised to lure in curious females: *come down and see my bower*. Does such obsessive courtship behavior really illustrate "male quality" the way hardline evolutionists want us to believe? Or do they show that evolution, over thousands of years, is able to produce art if there are no serious predators around?

The same questions can be asked about the extensive song of humpback whales, who have had millions of years to explore what can be created in the sea. Whale music may be the most efficient form of expression for so giant an animal. But like human music, it may express nothing but itself.

What would it mean to "know" a whale's song from the human point of view? To know the bee dance is to grasp what it would mean to need to dance out information, to feel the knowledge of the location of the flowers and the sun. To know the humpback song is to feel its resonance and its power, and to fathom a reason for its shape and form.

It is unlikely that humpback song contains as much practical information as a honeybee's dance. Whale music, for all its complexity, is probably much more like birdsong, whose beautiful patterns of sound elude explanation by simple function. After decades of listening and diagramming, we still have more questions about the song's structure and function than we have answers.

What biologists know about the humpback whale has unfolded gradually. We know the song consists of long repeated phrases that rhyme, because they end with similar syllables. It is the longest expressive vocalization of any animal; a single performance can last up to twenty-three hours. We know it is most often sung by solitary singers, always male whales, and that the females show little visible interest in what the males sing. But the singing most often happens in winter breeding grounds, so it is still believed to have something to do with mating behavior.

The best places to hear humpback whale song include scattered island locations like Maui in Hawaii or Archipelago Revillagigedo off Mexico for North Pacific whales, and Tonga for South Pacific whales. In the Atlantic, go to the Silver Bank off the Dominican Republic in the Caribbean, or the Cape Verdes off Africa. In the Indian Ocean, try Madagascar or Mayotte. In the Arabian Sea, go for Oman. In summer in the northern hemisphere, prime whale-watching season, humpbacks do not usually sing, but sometimes they surprise us and make music out of season.

The whales in each ocean have distinctly different dialects of song, but the dialects are not local: in any one ocean in any one hemisphere, humpbacks thousands of miles apart sing similar songs that change over the season in a similar way. No other animal develops its music so rapidly and regularly, and we have no idea how they manage it.

We do not know what, if anything, the song is saying. We do not know why it is so complicated and methodical. We do not know who is listening. We do not know why such changes from year to year are necessary. Whale songs may be changing for the same reason there are always new hits on the charts, because fashion doesn't like to sit still.

Science has been on the case of humpback whale song for four decades, but Payne and McVay's 1971 paper in the journal *Science* is still the most straightforward account of what is so special about this animal's music. In the opening paragraphs they announce something that no other scientists have seen fit to say in any technical paper, before or since: "Humpback whales emit a series of surprisingly beautiful sounds."

Too many whale scientists consider beauty to be too subjective to trust that term to describe the sounds they spend years studying. We must give Payne and McVay credit for not being afraid to express what they felt. This paper is the best one for non-specialists to read because it is the only published scientific paper that includes illustrations of entire songs, showing the whole shape of the sound. Read figure 17 like a musical score for a solo performer, from left to right, line after line.

After Payne handed over the tapes to McVay, he took to the basement of a Princeton University laboratory where sonograms of short sound fragments were laboriously produced by the rather primitive Kay Sonograph. Then Scott and his wife Hella traced the patterns by

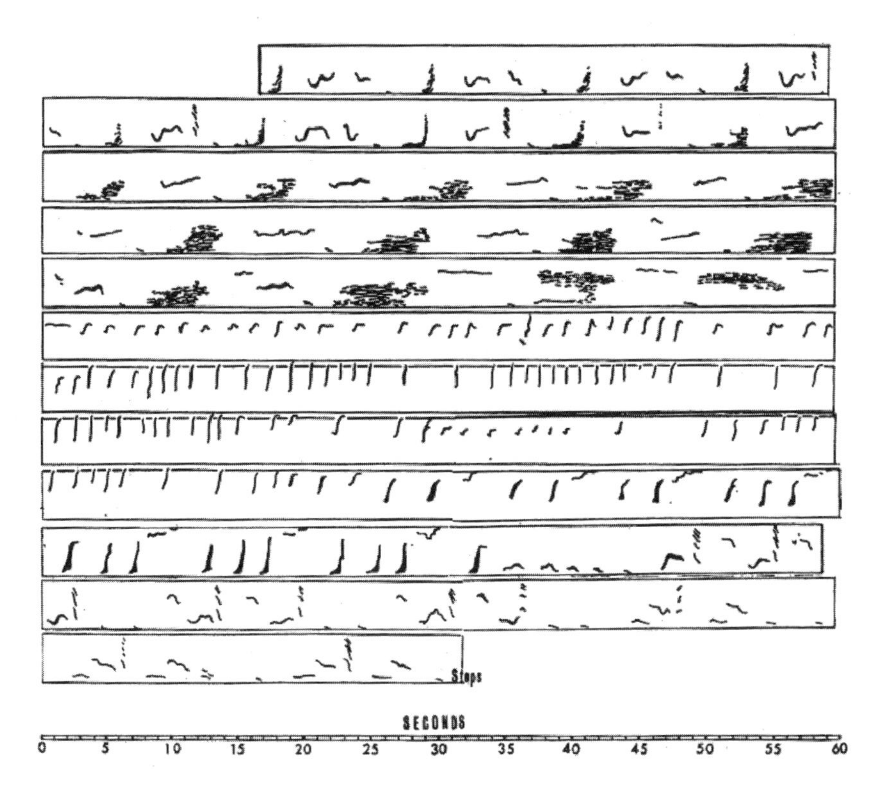


*Fig. 17. One complete humpback song, as transcribed by Scott and Hella McVay.*

hand onto long scrolls, spending hours sketching the kind of score that a computer could spew out in minutes today. They spread these alien hieroglyphics out on their living room floor and only then realized how systematic the whole song is.

This is a medium-length song, with the whale going on for about eleven minutes before taking a break. The sonogram allows a machine to notate sounds that can't be written in regular musical notation, revealing pattern and form. The horizontal tones are sustained, pure pitches, the lower ones more like moans, the higher like whistles. The big dark areas are washes of noise, many frequencies all at once. The vertical lines are rhythm-like pulses, the tall Ss are rapid upsweeps. A whole range of syllables are presented, with principles of organization visible at once.

This second song, recorded in a different season, shows how the whales combined the same elements in a somewhat different order in figure 18. It lasts for ten minutes:

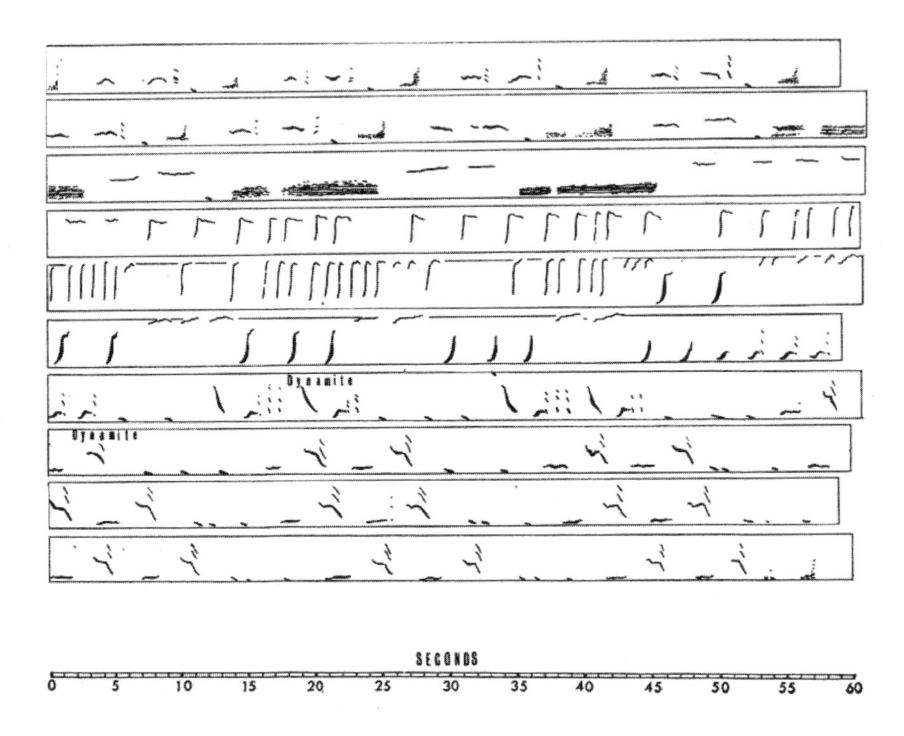


*Fig. 18. Another complete humpback song transcribed by Scott and Hella McVay.*

Why doesn't *every* paper that discusses the structure of humpback song include pictures where the songs dance across the page in a visual, easy-to-grasp way? I suspect scientists find the task too personal and artistic. The typical analytic approach gives every phrase an arbitrary name like "A," "B," or "C." Numbers are also human conceits, of course, and science is based on the idea that nature is a book written in the language of mathematics. It's a shame they forget that nature offers up beautiful music as well.

Payne and McVay identify three different kinds of sounds in figure 19: (1) rapid pulses alternating with sustained sounds, (2) short high upsweeps, and (3) low, sustained notes with complex timbres, often repeated over and over again. They are sung together in a tightly organized manner, with structure appearing at several hierarchical levels.

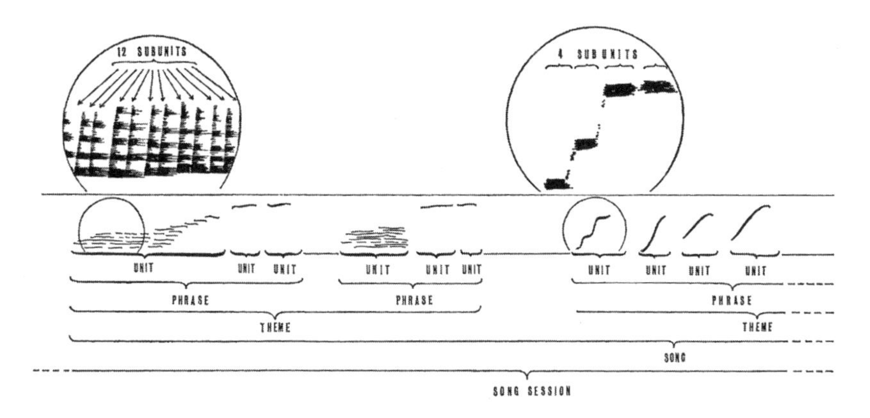


*Fig. 19. Hierarchical levels of structure at work in the song of the humpback whale.*

Each distinct sound type is named a "unit." These units are combined into regularly recurring phrases. Each phrase is put together in organized groups called "themes," and the whole ordered series of themes is called a "song." A single song can range from seven to thirty minutes, and a session of songs could go on for many hours, even all night.

This kind of structure for a complex song composed of repetitive parts has also been found in a few species of birds, most notably the nightingale, which has been extensively studied in Germany. Humpback whale songs are far longer and stretched out than the birds' buoyant motifs. The whales may have a much more languorous perception of time than we can fathom. Whereas birds have been tested to perceive short changes in rhythm about twice as fast as we can make out, whales may consider a seven-minute song to be quite short, with a thirty-minute series of songs just an average length, perhaps twelve hours really something.

Speed up a humpback song and it does sound a bit like a bird, not only because of similarity in structure but also because of the contrasts between sound units: birds also have sustained whistles, rhythmic chirps, and noisy *brawphs*. Similarly, slow down a bird song that features pure tonality and *rawps* of noise, say a catbird, and you will get something clearly whale-like.

Science strives for rigor and detachment, a move away from human bias and the feeling of what one hears. Yet how can science be objective when characterizing such alien music? Are scientists any more objective than the astonished music critics of 1970, flummoxed at having to describe to their readers what a whale song sounds like?

Payne and McVay were the first to try, and still the best. Let's go through each of the six themes they identify as part of "Song Type A," of which both of the complete songs above are examples. There are parallels in organization as well as variation—similarity, structure, and difference all at once. The six themes in Song Type A are usually heard in order: themes 1, 2, 3, 4, 5, and 6, although each time it is heard there is some variation. Here we go: theme 1 consists of phrases of three units, beginning with a sound like a motor rumbling, followed by clear, sustained moans, first warbling, then dipping, then steady, each time a bit higher, as shown in figure 20:

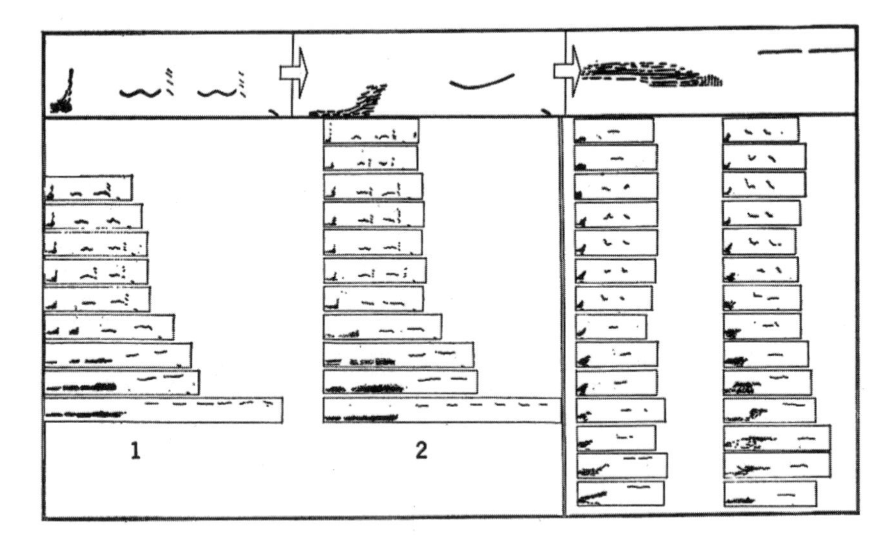


*Fig. 20. Theme 1 of Payne and McVay's humpback song.*

The whole theme shows a three-phrase structure that builds toward a conclusion with two high, sustained tones. Beneath the summary version are all the individual variations in sound heard during the same season. Sometimes there is a low grunt at the end of the first two phrases, sometimes not. (Whales are individual beings, not machines. They interpret the composition as they perform it.) The individual phrases noted down below show all of Payne and McVay's raw material, so you can see how they generalized from the different variations they heard.

Figure 21 shows theme 2, a long rhythmic series of piercing *buweeps*:

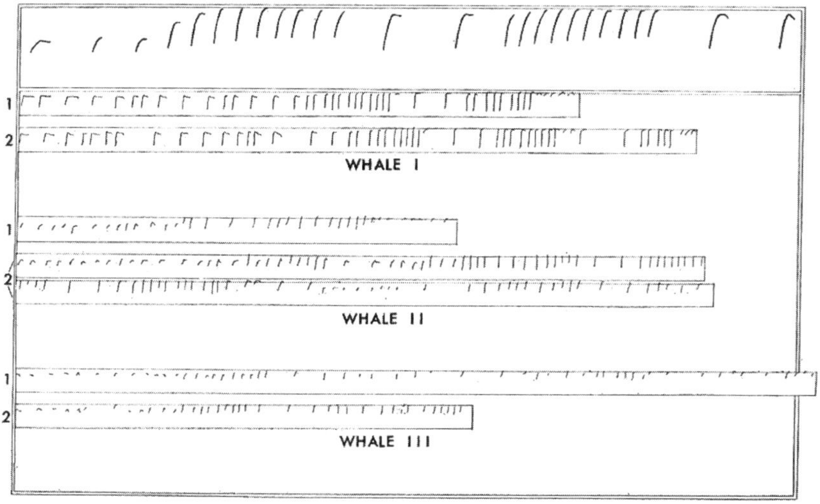

*Fig. 21. Theme 2 of Payne and McVay's humpback song.*

Like tentative hieroglyphics or the beginnings of visual language, this is the most variable theme, with the upsweeps sometimes giving way to quick, high chirps.

Themes 3 and 4 in figure 22 have the great sweeps morphing into low, slowly rising pure tones. In between the sweeps are astonishingly high cries:

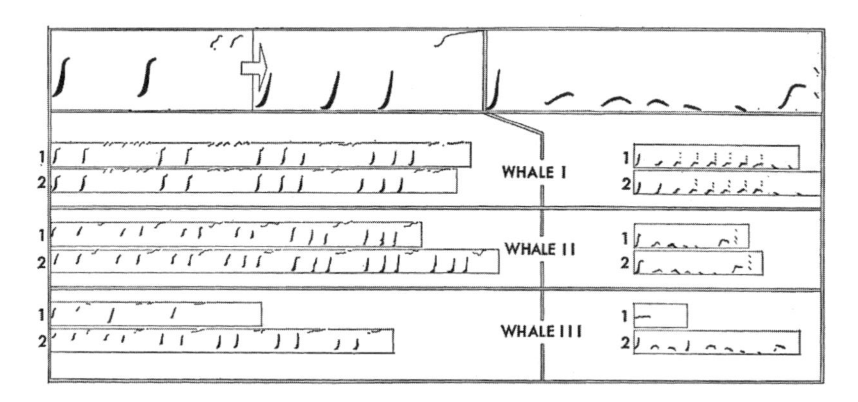


*Fig. 22. Themes 3 and 4 of Payne and McVay's humpback song.*

Theme 4 seems to end with increasingly strong, deep bellowing units. There is less variation in these later themes so Payne and McVay don't show so many versions.

Themes 5 and 6 are evenly spaced, rigid in time, ending with two low grunts. The main difference between them is that in theme 6, the low tones are steady and hold to a single low pitch, while in theme 5, they wooze and waver in figure 23:

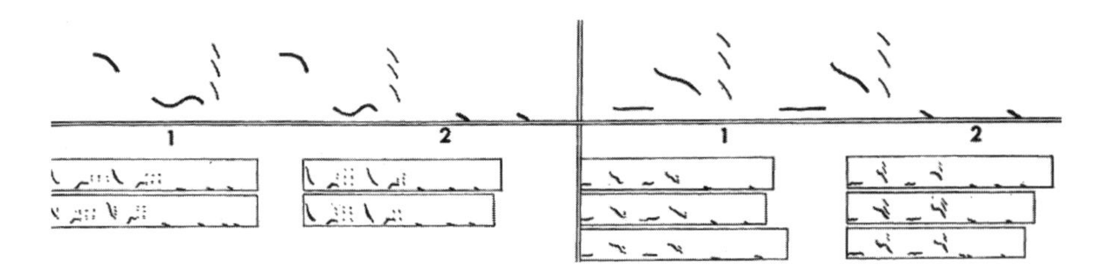


*Fig. 23. Themes 5 and 6 of Payne and McVay's humpback song.*

The traced sonograms of humpback whale song resemble the notation of Gregorian chants written down in the tenth century. These fluid musical notes are called "neumes." They did not represent exact pitches or rhythms, but tendencies of patterns, exactly what McVay and Payne identified in the songs of the whales.

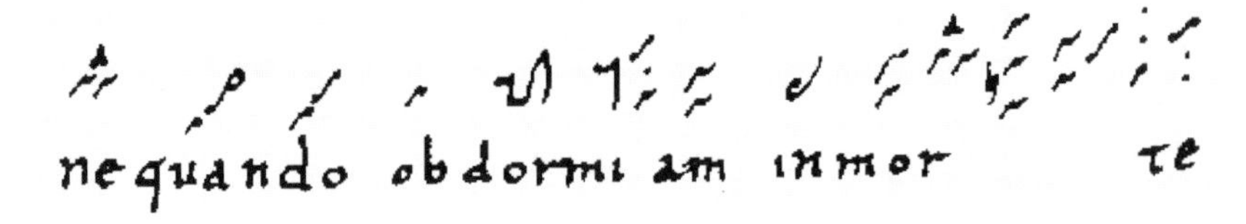


*Fig. 24. Neumatic musical notation from the tenth century.*

From the tenth to the fourteenth centuries these neumes evolved slowly into the musical notation we know well. As the tuning of instruments standardized, and music evolved to move between one key and another, it became increasingly important that the pitches of notes be indicated more precisely. It took hundreds of years to standardize human musical notation, so who knows how long it will take us to develop an official method to write down whale sounds?

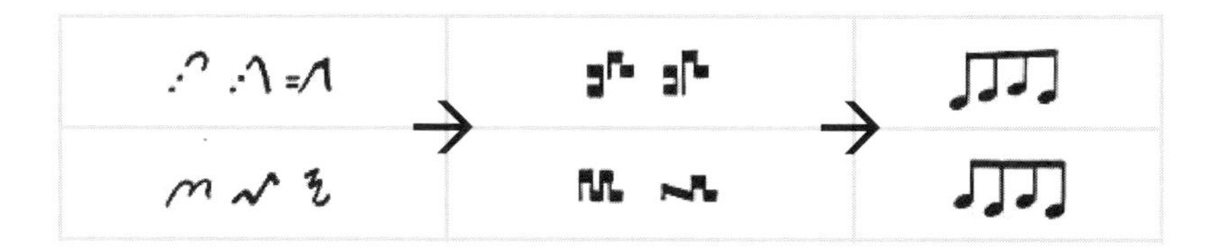

*Fig. 25. How musical notation got more precise from the tenth to the fourteenth centuries.*

Looking at McVay and Payne's neumes on the page, we can immediately grasp rhythm, shape, and form, even if what we hear is too slow to make sense of. Peter Tyack credits this clear graphic presentation of the humpback whale song structure as the first thing that got people to take its structure seriously: "This is the wonderful thing about a scientific discovery. A phenomenon is carefully described, then it is suddenly part of our lives. But I don't think anyone predicted the huge emotional impact whale song would have." The qualities of the song are so remarkable that biologists could not ignore it. Three more descriptive aspects of the song were discovered in the 1970s and 1980s that led biology to take even greater interest in whale vocalization.

First, Katy Payne and Linda Guinee pointed out that because each theme ends with consistent final sounds, the phrases could be said to "rhyme" in a way akin to human poetry. Different patterns in each line end with the same sound, making the larger level of rhythm easily recognizable, to us and presumably to the whales. Thirty-six percent of phrases in the themes, on average, end with a rhyme-like unit. No bird does anything like that. Payne and Guinee published their results in 1983.

Second, the song is performed only by males. To accurately sex a whale, scientists shoot a small dart into the whale's back and remove a little piece of tissue. (Though such techniques might sound like harassment, scientists claim it hurts little more than a mosquito bite.) DNA analysis tells us what chromosomes are there, so it was quickly determined that only the male whales sing, and most often, a solitary whale, making music all alone. The songs are sung mostly in the winter, in the places where humpbacks gather to mate. (Although Bermuda, where the song was first documented, is not one of those places—those whales were migrating through on their way to summer feeding grounds in the North Atlantic.)

After the first decade of study, it seemed likely that the song of the male humpback whale is designed, like bird song, to attract female whales, or possibly fend off unwanted competition from other male whales. Only one problem: over several more decades of observation, no one has ever seen a female humpback whale approach a singing whale. They are simply not interested.

Are we missing something? Perhaps the whales are like us, flirting during the day from afar, getting serious only at night when scientists are not usually watching or listening. Or are we blinded by science, misguided by our own most excellent but possibly inappropriate theory of sexual selection?

The third major discovery about the humpback whale song is its biggest challenge to classic sexual selection: at the beginning of each breeding season, all the male whales sing roughly the same complex song, with its series of themes sung in order with only minimal variation. As the season progresses, their song changes rapidly, from week to week.

And the following season, it changes again. Ten years down the line, it's a whole new song. So what Payne and McVay have analyzed for us, and what Judy Collins and Paul Winter have played along with, is sung no more in the wild. It exists only in old recordings and diagrams! Whereas bird song is the oldest animal music we know, sung consistently for millions of years, humpback whale song is the newest, showing animal culture in action at breakneck speed, changing from week to week.

Some say the songs tend to lengthen from year to year, as illustrated in a paper by Katy Payne and Peter Tyack in the 1970s. "The song changed drastically during that period," said Tyack, "but the interactions hardly changed at all. The change then doesn't reflect changes in message, but is more akin to changes in musical style." Tyack and Katy Payne listened to humpbacks in Hawaii for five seasons in a row, and this is what they found, shown in figure 26:

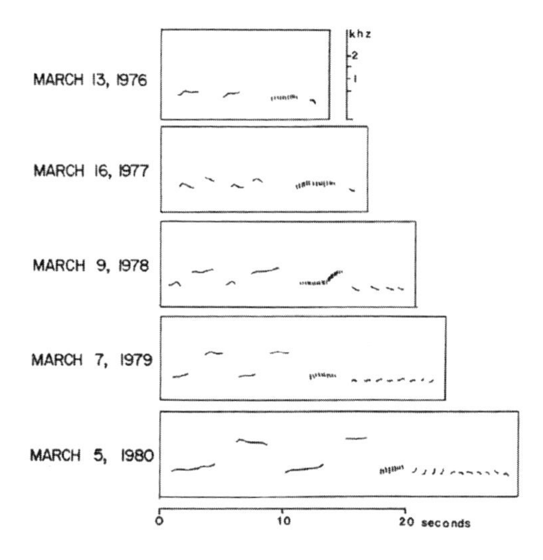


*Fig. 26. A portion of the humpback song lengthened from year to year.*

More remarkable still is how much the themes can change within a single season, as shown in figure 27, in parallel ways among populations separated by thousands of miles. How can singing whales off Hawaii and those off Mexico, who never meet during the breeding season, change their songs in a similar way over only a few weeks, when they are too far apart to hear each other?

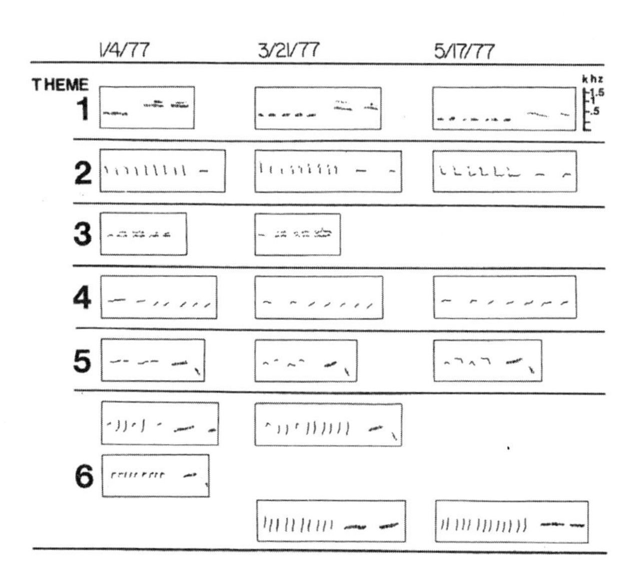


*Figure 27. How humpback songs change from month to month in a single season.*

Tyack believes there are probably enough whales evenly spaced across the ocean from the Hawaiian to the Mexican breeding grounds to transmit the song rapidly across the seas. "We haven't proven there's a mystery," he cautions, "but the greatest puzzle to me is how *similar* the songs are at any given time from whale to whale."

Roger Payne loved those 1960s songs so much he lamented their passing: "Today's humpback whale songs pale beside those of the sixties," he recently wrote in an open letter to the youth of Japan. "The North Atlantic is so musically lackluster today." I don't think Paul Knapp would agree. Today there might be more beats, maybe fewer legato passages. We may like the beat more than melody today, and it might be the same with the whales.

Except nature is not supposed to work that way. Or if it does, it usually happens very slowly, over thousands or millions of years that allow a species to evolve its salient aesthetic qualities.

There are a few exceptions. We know of two species of African songbird in which the males change their songs over the course of one mating season. This behavior is surprising enough so that both the yellow-rumped cacique and the village indigobird have been investigated in some detail by ornithologists. These birds live in small colonies of twenty to fifty individuals, and male birds are able to learn new songs even as adults, a skill not possessed by most songbirds. Males who sing certain songs enjoy greater mating success. When less successful males imitate those songs that are popular with the ladies, their chances of mating improve. You've got to know the right tune at the right time.

How do the dominant males stay on top? Once they see that all the other males have mastered their hit song, they start to improvise, and soon create a new song that sets them apart once more. Then the other males copy again, and the cycle keeps going, very rapidly, from day to day and week to week over the breeding season.

Each colony has its own song variations, which some have called "dialects," a word I don't like so much because it makes all this musical one-upmanship sound more like language. To me it's closer to a musical competition, and as a jazz musician I like the fact that it values improvisation. The females favor not only the males with the best tunes but also those males who come up with new melodies once a tune becomes too tired: you don't want to be known for the same old licks.

Classic Darwinian sexual selection, but with the twist of rapidity. Female preference drives male innovation, not over the thousands of years it takes to define the salient character of a species, but from day to day over one season. If you listen carefully, you can hear evolution at work in the course of a single week.

Is this the way it works with humpback whales? There are definite similarities, first documented by Sal Cerchio, Jeff Jacobson, and Tom Norris. They spent the summer of 1991 listening to eleven whales off Kauai and thirteen off Mexico, separated by thousands of miles of open ocean, but all singing the same song, a song that evolved in parallel over that one season. Then it took the three researchers *nine years* to process all that data and come up with enough certainty to publish their results in 2001.

Over that one three-month season, the same specific changes occurred in the songs of both these populations. For the same competitive reasons as in the indigobird and the cacique? Can there be a sort of "cultural transmission" where the whales learn from each other, even over thousands of miles of open ocean?

Cerchio and his colleagues point out that there are significant differences with the humpback song story that suggest something else was going on. Why did the Hawaiian and Mexican whales sing the *same* changing song, with no regional dialect differences? Humpbacks do not live in small colonies where such rapid behavioral change can be easily explained—thousands of whales separated by such great distances can't really be competing for the same females.

Can the songs themselves travel so far? Only the lowest-pitched intensely slow whale beats could, in theory, become thousand mile songs echoing across leagues of sea. Only a few of the humpback's tones are deep enough for this kind of travel. Humpback tunes, drawn-out as they are, are still too high and fast to reverberate their precise turns of phrase more than ten or twenty miles under the water.

Cerchio's paper proposes that there might be an "inner template" for structured change inside the mind of the whale. Even all alone, a male humpback will sing, and then gradually modify his song over the course of a single season in varying, but somewhat predetermined ways. Cerchio says the phrases tend to lengthen over a season, and hardly ever get shorter. The tendency to lengthen a phrase could be some kind of song-guiding rule, an inner template in the brain of the whale. It would take much more data, and many more seasons of watching and listening, to even begin to know. The authors admit that such an idea, based on the twenty-five whales they were able to listen to in 1991, is sheer speculation.

Over the four decades since Watlington turned his tapes over to the Paynes in Bermuda in 1967, thousands of hours of humpback whale song have been recorded. Much of this sound still awaits analysis, since the process is so time consuming, whether done by hand on the living room floor, or sifted through number-crunching algorithms on the latest computers. It's easy enough to spew out sonograms, much harder to figure out what to do with them.

When I spoke with Cerchio at his office at the American Museum of Natural History, he maintained that there is more to the story than he was able to present in a purely scientific paper. "Few people have put out solid hypotheses on why whale song changes." He's a thoughtful guy with years of experience and many hunches he has yet to try out. "I can imagine an ancestral song being simple, not changing, with selective pressure to increase the complexity of the song, a female choice mechanism for complexity, which has real physical ramifications. Humpback whales are capable of making a broad range of sounds. Their vocal mechanism has to be far more complex than fin whales and blue whales, who have very different selective pressures." It's all a question of the unknown mechanics of evolution. "Whatever selective force they had allowed them to develop a wide frequency range, great variability in harmonic structure, and the ability to do things that other whales can't do. Change in song came secondarily after that, when you had complexity tapping out at the limits of their ability."

Cerchio is a firm believer that sexual selection is the means that drives the humpback whale's unique behaviors, a proven process responsible for much of the most puzzling kinds of diversity in the animal world. He's not troubled by the fact that most accounts say female humpbacks show no reaction to the males' songs. "Oh, we've seen it," he assures me. "Everyone's seen it, but if you have a different hypothesis, you'll tend to ignore it."

So why would an animal need to change its song so rapidly, over so wide an area as an entire ocean? "At any given moment, you look at the songs of two individual humpbacks, you can tell them apart, but over time, the temporal variation swamps individual variation. The fact that they're changing together is more noticeable than the individual differences over a single season."

"If you can't tell the individual whales apart, how can there be sexual selection on the basis of the song at all?" I wonder.

"That's a good point. The cue has to be the change itself! It must be an indicator of some kind of male quality that the females can perceive."

If you believe that sexual selection explains everything, you will see it at work everywhere. It's one thing most biologists take on faith.
Only rebels in the field will claim that humpbacks are up to something else.

But those rebels are the people who've been listening the longest.

As the world's largest island, Australia has two populations of humpback whales, one in the Indian Ocean, off its west coast, and the other in the Pacific, off its east. For most of the 1980s, Douglas Cato listened to humpbacks off the east coast for the Australian military, working for one of the longest-running whale-listening programs on the planet. Cato liked to describe the differences between themes in terms of the raw qualities of the sound, using more colorful language than American scientists: he heard things like moans, sighs, violins, gulps, yaps, chirps, and chainsaws. In 1991 Cato's associate Astrida Mednis coined the term *screal* to describe that sound somewhere between a scream and a squeal that so often ends the theme, giving that sense of rhyme. The sound qualities of the humpback's song suggest whole new words that lie at the very the limits of our ability to describe.

In 1996 biologist Michael Noad took over Cato's east coast project and set up a series of hydrophones and recording devices fixed to buoys that could record the sounds of the whales twenty-four hours a day. By the end of that year he recorded a total of eighty-two different whales, almost all of them, as expected, singing the same song. But two of the whales sang a different song, a completely different arrangement of yelps, booms, and screals. Why? Who were they? Where did this new song come from?

The next year, Noad and his team returned to Peregian Beach, just south of Queensland's Sunshine Coast. This time, a third of the whales were singing the song that was so rare the year before. This was unprecedented in humpback whale song research—one population of whales usually sings the same song at any given time. What was happening here? By the end of the 1997 season, the new song so gained in popularity that the old one nearly disappeared. This kind of "cultural evolution" in whale song was not supposed to happen *so* fast.

At first Noad assumed the new song had emerged spontaneously, like some surprise mutation that suddenly started to catch on. Then he decided to call on his colleagues on Australia's west coast. Previous studies had shown little similarity between the songs of each population, since they are usually separated by thousands of miles of land-mass. But the moment Noad heard the 1996 west coast song, he knew this was the same aberrant song that had appeared that summer on the east coast. Now he assumed that a few Indian Ocean whales had gotten lost, a few thousand miles east, and introduced their song to the Pacific population.

But why would the eastern whales immediately drop their own song and pick up the new one—just because it was new? Cerchio suggests that humpback whales may crave novelty for its own sake, evolving in males a natural desire for innovation, like the village indigobird, whose song also changes over a single season, as all males in a local population pick up the changes. Those males that have the highest mating success are the ones most listened to by other birds. They're the birds who lead the innovation in the song.

Cerchio wants to combine this tale of ornitho-musicological innovation with normal models for terrestrial mammals. "The rate of change is likely related to population density. Where there are many more whales the change will come faster. Secondly, it may be related to immigration, not just two whales like Michael Noad suggests, but a *pulse* of individuals."

With the sudden appearance of full-blown, brand new humpback songs in their midst, the Australian Pacific whales were impressed enough to quickly take on the new tune. Noad thinks this is because the new tune appeared "above a certain threshold level," meaning that it was loud, consistently present and clear enough to suggest a real alternative that the

Pacific whales hadn't tried before. An entirely new song, perfectly formed and booming out of the sea, is something that hardly ever happens.

This suggests, to the musician in me, that aesthetics plays a serious part in the evolution of humpback song. Not aesthetics in the sense of what music is good and what is bad, but in the way evolution produces a sense of *rightness*, from each species' perspective, about how animals should look and how they should behave. The desire for the new in humpback whale song is at the heart of this aesthetic sense. Yet we have to wonder why this Australian Pacific Ocean song changed so much faster than usual. Was the Indian Ocean song that much better to all the whales who heard it? Or was it just something new and different?

Noad was fortunate to be listening during the very season that the eastern humpback song changed the most. But his overall explanation is rather pedestrian: if novelty really is the driving force for song change, it implies that the primary purpose of song is to attract females, since it's only through sexual selection that the new arrives in appearance or behavior for its own sake. Still, if males really are singing for females, we ought to see females show some interest in the music once in a while. Perhaps there is more to the story.

What is the best way to grasp the structure and pattern of these songs? Payne and McVay wrote them out longhand, so they resemble musical scores. Other researchers focus instead on categorizing the different types of sounds, and then tabulating how often they appear and in what sequence. Here's how Whitlow Au, Adam Pack, and Louis Herman tried to do it. They heard nine basic kinds of sounds, as shown in figure 28.

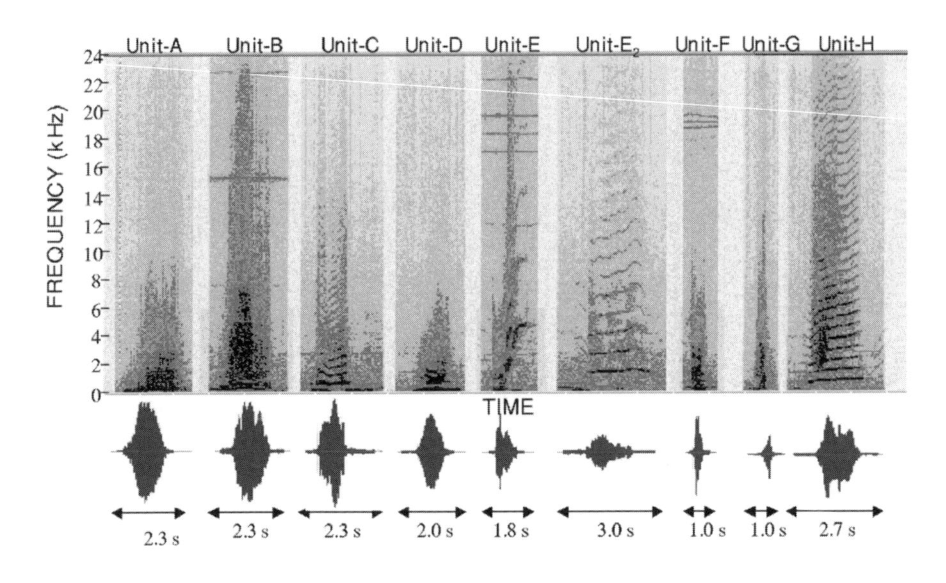


*Fig. 28. Spectrogram and waveform representation of the nine basic units contained in the humpback whale songs during the winter season in Hawaiian waters in 2002.*

That's how the computer printed them out, and here's how the scientists described them in words:

A—vibrating upsweep
B—double upsweep
C—frequency sweeping cry
D—flat tonal groan
E—low gulp jumping to very sharp upsweep
E2—high-frequency tonal wail
F—short low-frequency downsweep
G—short low-frequency upsweep
H—mid-frequency tonal wail

Far more technical than the language of the Australians! The team then characterized four different themes composed of these units and began to grasp an outline of the structure:

The general sequence of units producing the four themes observed in the recordings of nine singers. There is considerable variability in a theme associated with the number of or sequence of the units. The units in parentheses are those that are repeated seemingly randomly from one to seven times, depending on the specific whale.

| Theme | Units |
|---|---|
| 1 | B C (B-C) D D A |
| 2 | D A D E F D (E-F-D) E A |
| 3 | A F F F A F F E F D (E-F-D) E2 E |
| | A F F F A F F F F E E E [without (E-F-D)] |
| 4 | G G G G H |

The specific number of F units can vary between 2 and 7, and E units between 1 and 3. Sometimes the H unit is not emitted.

*Fig. 29. How the nine units are put together.*

Here we have a more schematic way of representing the songs of the humpbacks than the wispy scores of Payne and McVay. These more modern sonograms, ironically, give much less of a sense of the shape of each phrase and its place in the song.

There are clear qualities that every humpback whale song has: a specific range of kinds of sounds, and a specific use of rhythm, similarity, and contrast. The precise balance of these elements has mostly been described in statistical ways. That is why I think a musicologist ought to have something particular and interesting to say about the aesthetics of the song: what makes it uniquely humpback, rather than say, bowhead (the only other whale that sings anything remotely structured like a song).

Dividing music into its parts and forms is no substitute for the experience itself. The humpback song itself is still more complicated because of the layers of organization, first noticed by Payne and McVay. Sound units are combined into phrases, which are combined into themes, like the four noted above. These themes are combined into songs, and then the songs are combined into song sessions. At what level does analysis yield the clearest results?

In another paper, Pack, along with Eduardo Mercado, tried to sidestep this hierarchical approach and pay attention instead to certain patterns that seemed to occur much more often than others:

| | | |
|---|---|---|
| UNITS | = | {A, B, C, D, E, F, G} |
| SUBPHRASES | = | {DE, FF} |
| PHRASES | = | {BC, DEDEFF, GGGGGGG, AAAA} |
| THEMES | = | {BCBCBC, DEDEFFDEDEFF, GGGGGGG, AAAA} |
| SONG | = | AAABCBCDEDEFFDEDEFFGGGGGGG |
| SONG SESSION | = | ..A..BC..DEDEFF..G..A..BC..DEDEFF..G.. |
| SOUND PATTERNS | = | {A, BC, DEDEFF, G} |

Representation of the types of structural components typically present in sequences of sounds produced by singing humpback whales. Each letter represents one sound. Each individual sound is called a unit (different letters correspond to aurally distinctive sound units). Repeated groups of units are called phrases. Some phrases consist of repeated groups of subphrases. A theme is a set of repeated phrases. Songs consist of repeated theme sequences within a song session. We use the term 'sound pattern' to refer to any sound or set of sounds that is consistently repeated within a song session.

*Fig. 30. Pack and Mercado's hierarchical structure.*

They hypothesized that some parts of the song might have a different function than others. Perhaps the low sounds are meant to carry farther across the ocean, while the higher sounds are for close-range communication. Mercado has long argued that the songs of humpbacks are not for mating (because of the lack of evidence) but instead might function as a kind of sonar to help the whales navigate and locate each other. The repetitive sounds in particular might have such a function.

Although he wrote one of the most detailed dissertations on the song of the humpback whale, defending this hypothesis with detailed engineering and underwater sound flow diagrams, it has found few adherents in the field. The documented sonar sounds of marine mammals are more often sharp, quick clicks and creaks. It is unlikely that such a carefully shaped song could have a function so far away from its form. Plus, could it be true that only the males need sonar, not the females? Highly unlikely.

There are other left-field hypotheses, mostly without evidence. Chu and Harcourt thought the long, drawn-out tones show the females how long a male can hold his breath under water. Hafner and Winn surmised that the cry-like part of the song has enough individual variation among whales to be a kind of signature whistle. Again, these are intriguing hypotheses for which there is little or no evidence. "Why is it," wonders Cerchio, "that in marine mammal science there seems to be a willingness to accept a hypothesis before we have any means of testing it?"

All these approaches step away from hearing the song as a patterned whole and toward imagining that it might be a mix of sounds whose purpose is constantly varying, more like the interactive noises of belugas and orcas. But even this hypothesis offers little explanation for why and how the song changes regularly from year to year. Noad at least heard a few new whales in his population, and he documented how the new tune caught on, far more rapidly than anyone thought possible.

A study of the whales surrounding the island kingdom of Tonga in the South Pacific shows that the change in how themes are organized into songs is far from random. Nina Eriksen and colleagues from Denmark investigated the songs of southern hemisphere humpbacks throughout the 1990s, around the same time as the appearance of that new Australian song. These whales migrate from Tonga to the Antarctic and back, much like the Hawaiian humpbacks head north to Alaska. They sing most in the months of June and July, which is winter in the southern hemisphere. The same general patterns of cultural change were observed. At least one new theme appeared every year except 1996. Unlike Hawaiian songs, where existing elements tend to lengthen and get more ornamentation, some of the phrases in 1998 seemed to shorten. Eriksen believes this contradicts Cerchio's hypothesis that the whales might have a simple, inner template to govern how each phrase changes. She found more innovation and variation in both directions; shorter and longer, slower and faster.

Figure 31 shows the basic sequences of themes (not phrases) that they heard from year to year, making up the full song:

| Year | Median theme sequence | | | | | | | | | | | | | | | | | |
|---|---|---|---|---|---|---|---|---|---|---|---|---|---|---|---|---|---|---|
| 1991 | | | | | | | | | | | | | | I | J | K | L | M |
| 1993 | A | | | | C | D | E | F | G | F | G | F | | | | | | |
| 1994 | A | | B | | C | D | E | F | G | | | F | | | | | | |
| 1995 | A | A2 | B | | C | | | | G | | | F | H | | | | | |
| 1996 | A | A2 | B | | C | | | | | | | F | H | | | | | |
| 1998 | A | A2 | B | C2 | C | | | | | | | | | | | | | |

*Fig. 31. Eriksen's account of annual humpback song change in the South Pacific.*

Each theme follows the next in the same order, as noted, with a few exceptions. To the Danes it made sense to think of the music as a circle, not just a table, as in figure 32. This kind of diagram was first pioneered by Katy Payne in the eighties:

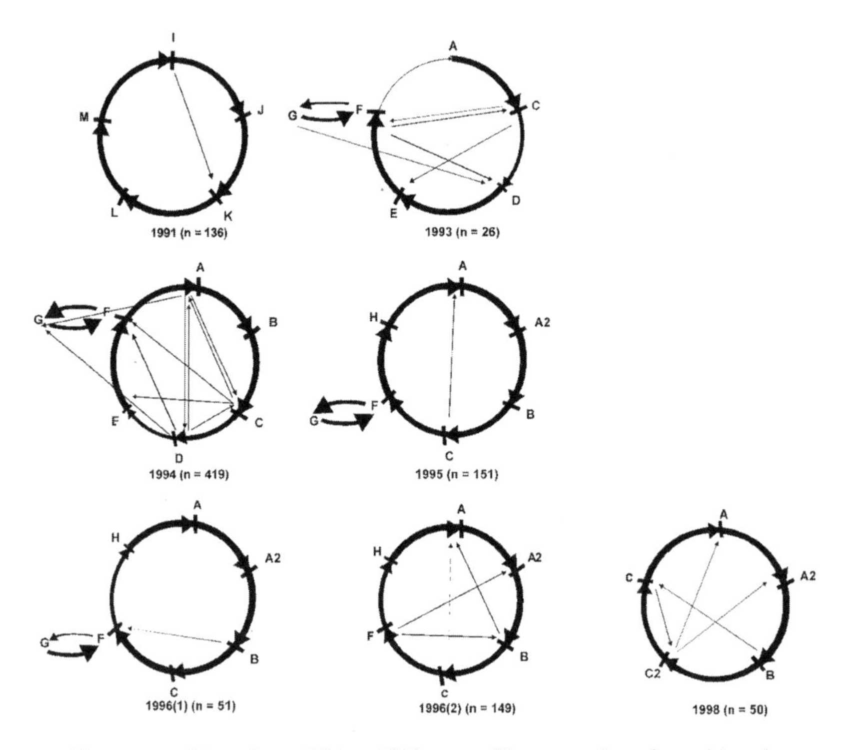


Themes transitions from 1991 to 1998 songs. The proportion of transitions is indicated by the thickness of the arrows. The number in the parentheses shows the number of theme transitions of each year. Note that 1994 had the most variable theme sequences and the largest number of transitions.

*Fig. 32. Eriksen's circle of humpback whale song themes.*

Looked at this way the song seems to have no beginning or end, representing a great cycle that all the whales in a given year must know.

Figure 33 shows how just one of the phrases in one of the themes lengthened from 1995 to 1996, and then got shorter by 1998:

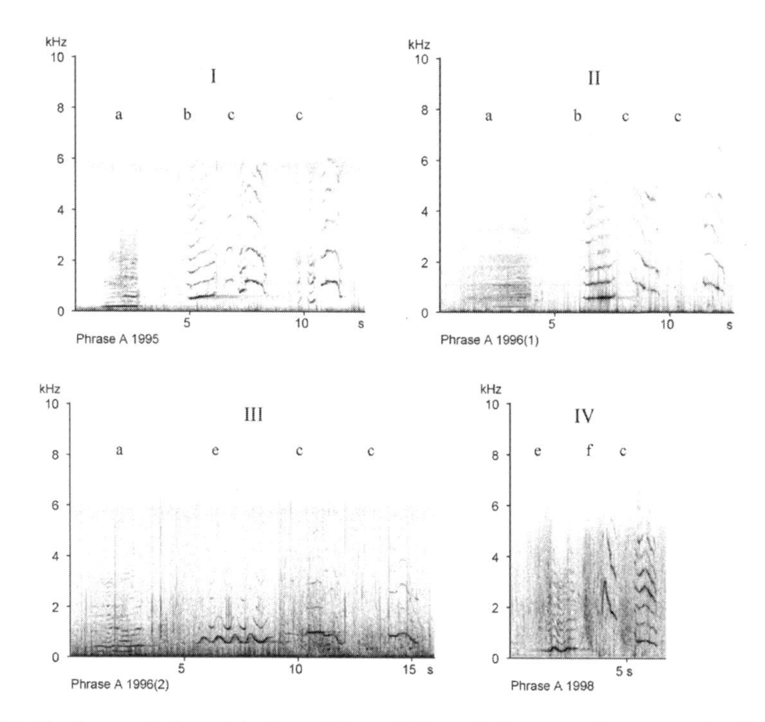


*Fig. 33. The changes of phrase A in theme A from 1995 to 1998 in Tonga. Phrase A consisted of three to four units: phrase structure was retained from (I) 1995 to (II) 1996 (1). In 1996 (2) Unit B was substituted by unit E (III), which made the phrase last longer than in the previous years and 1996(1). Unit C was still repeated twice in 1996, but in 1998 the last two units were not similar (IV). In 1998 only three units were present, as in 1995. Over the three years, one humpback whale phrase got longer, then it got shorter.*

It seems to take just one year for a song to spread through a population in this part of the South Pacific. Eriksen believes that the changes are learned by one whale from another, not invented in each whale's mind, because two different songs were present in 1991 and 1996, possibly related to the differences heard that year by Noad in Australia a few thousand miles away in the same ocean to the west.

Cerchio doesn't believe Eriksen has read his results right. "I'm still seeking to test my hypothesis. Today I'm also working in the southern hemisphere. We're looking at one population of humpbacks off of Madagascar, another off Gabon. They sing remarkably similar songs. If you take those phrases that they share, I would predict in both regions, the progressive change would be similar. The easiest variable to measure would be phrase duration. You can immediately eliminate cultural reasons for the similarity, since the populations are too far apart to ever meet."

"If the phrases only lengthen wouldn't they just get endless?"

"They cleave into two." That's what happened with Nina's one phrase that got shorter. "In her paper, Nina Eriksen challenged my results. But six of her themes got longer, only one of them got shorter. I consider that overwhelming support for my hypothesis!" Sal beams optimistically.

In June 2007 Elizabeth Derryberry published a paper in the journal *Evolution* that brings surprising support to Cerchio's argument. It turns out that male white-crowned sparrows have also lengthened their songs between 1979 and 2003, showing for the first time that a bird species is changing its song in the same manner he has hypothesized for humpback whales. In the white-crowned sparrow case, today's females definitely preferred the up-to-date song to the hits from yesteryear. We have yet to find such confirmation from female whales.

There is still the question of how the change is passed on. "Didn't Noad prove that the change is learned from whale to whale, and that it can happen really fast?" I thought I understood that much.

Cerchio is not convinced. "Noad showed that the changes *are* culturally transmitted, but the *rules* that govern change are most likely genetic, as in birds." The brain of the animal must first contain an ability to change the animal's sound, a skill that few creatures possess.

Now Sal is starting to make sense to me. What he calls genetic rules I call the species *aesthetic*, the idea that the whales prefer a certain kind of song because it is "right," not because they need music to convey information.

Darwin wrote in *The Descent of Man* that birds have a natural aesthetic sense, which is how he explained the preponderance of beautiful plumage and songs. Biologists have tended to ignore this remark, but it remains a valuable insight, given the kind of data we're collecting today on whales as well as birds. Look at these complex patterns: How much information can be contained within them? Are the whales conveying specific and evolving information to each other? Despite the complex and evolving patterns in the music, biologist Peter Tyack and engineers John Buck and Ryuji Suzuki say *no*.

They used information theory to analyze the humpback song, because this is an approach that can look at a signal with no knowledge of its context and figure out the maximum amount of information contained within it. This is convenient because it sidesteps the question of what the song is actually for from the point of view of the whale. It's akin to the process of breaking enemy codes in wartime, trying to find meaning inside patterned gibberish without knowing the rules.

Tyack has been listening to whale sounds for years, but hadn't solicited the help of signal specialists until he met Buck and his student Suzuki. They took an easily quantifiable approach: How much stuff happens per second? This is the bit rate: for whales, from .1 to .5 events per second; in human speech, 10 to 20 bits per second; for computers communicating over the internet, 56,000 bits per second is a slow dial-up connection, and a fast connection is at least a million bits per second.

Does that mean machines are smarter and faster than everyone else? Only in terms of bit processing. Real communication requires a lot more than that. It needs that whole context that information theory so conveniently avoids. Animals, with all their rich range of sounds and behavior, are not known for explicitly sharing much information in their lives together, even in their most extreme performances. Birds of paradise in New Guinea dance and preen their shocking colors, but the purpose is said to be all the same, to impress the girls. The splendid ornamentation of bird songs cannot be explained by saying they all serve the same sexual purpose. Peacocks' tails are anything but useful.

Why are there elaborate songs then in birds and whales that convey no complex message at all? I asked Buck what would happen if you plugged a solo cello sonata by Bach into his schema, with all its levels of structure and repeating patterns. "Not much information there," he responded. What about the rich, screaming harmonics of Jimi Hendrix playing "The Star-Spangled Banner," another performance that some have likened to the screals of whales? "Not there either."

Wait a minute, is human music then about *nothing* as well? Only if you think of communication as made of information and nothing more. Music is important enough to have evolved along with humanity for at least hundreds of thousands of years. We may make music to charm the opposite sex, but only a self-satisfied biologist would say that reason for music is enough to explain it. We, at least, are one species who spend a lot of time playing around with sounds for their own emotional and beautiful qualities. Why not accept that other creatures could dwell in music the same way? Why couldn't it come earlier on the evolutionary tree, *before* language, in which the parts of an utterance mean something separate from the whole?

Archeologist Steven Mithen believes that our immediate ancestors didn't have language in the sense of complex syntax for complex messages. In *The Singing Neanderthal* he writes that our predecessors had a much more emotional sense of communication. They had a complex array of grunts, cries, and moans, organized in various patterns. These sounds were much closer to music than they are to language. Since all Neanderthals had was this emotive repertoire of sounds, Mithen surmises that music meant much more to our ancestors than it could ever mean to us—as music might mean more to whales than we can ever know.

I asked my friend Partha Mitra, who has written some of the signal-analysis software used in comparative bioacoustics, what he thought about the information theory whale song paper, which had been touted by the MIT press office as "finally cracking the code of the whales." Mitra said that as a mathematician, he knows that "information measures are often used for bad reasons, just because they sound sophisticated." He likens this approach to Pythagorean numerology, taking numbers as a kind of sacred tool to decode biological complexity. It's all dressed up with nowhere to go. Information theory ends up being more about what it is able to count than anything else, more information about information. Scott McVay read the paper too. He said, "These numbers and charts just confirm what we sketched out by hand long ago."

Even Peter Tyack, who wrote the paper, recognizes its limitations. He didn't mind that the whales off Hawaii were not telling elaborate tales to one another: "Animal communication is more complex and more interesting than just bits per second. If you consider each unit to be like a letter, then there is a high amount of redundancy, and a low bit rate. That's another way of saying that the song is highly structured, certainly not random. It's obvious to anybody who listens to humpback songs that they sound musical, and clearly there is similarity among the music made by all kinds of animals." It may depend on emotion, beyond any reason.

How could we know if great whales, with lives so far away from ours, experience emotion? In 2006, two researchers at the Mount Sinai Medical

Center in Manhattan, Patrick Hof and Estel Van Der Gucht, discovered that a particular kind of brain cell called the spindle neuron, previously known only in humans and great apes, was actually found in the brains of humpbacks, fins, and probably other large baleen whales. These long, drawn-out cells are found in the two parts of the human brain where empathy, intuition, and sudden gut reactions are supposed to happen. And they're found in exactly the same place in the brains of baleen whales, as shown in figure 34:

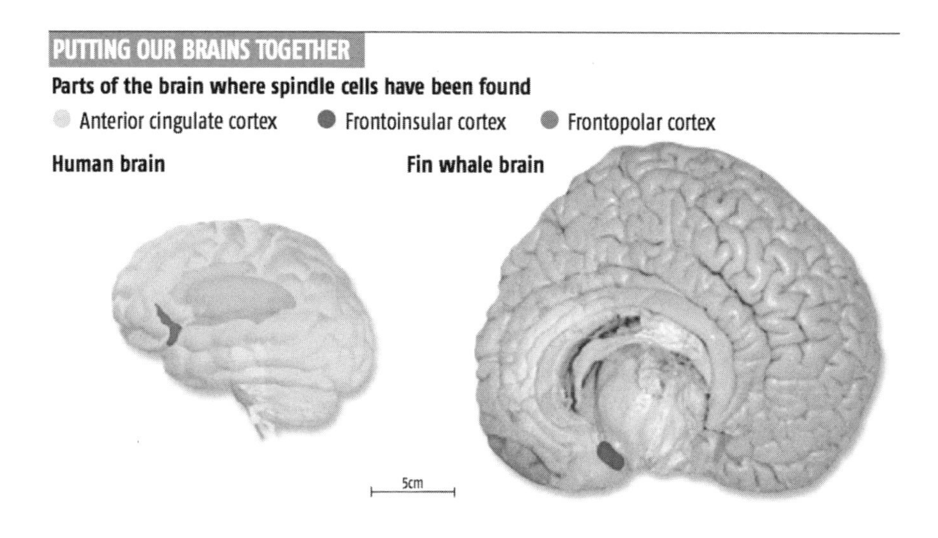


*Fig. 34. Where the spindle cells are in the brains of humans and large whales.*

In humans, the anterior cingulate cortex senses pain, admits errors, and focuses attention. Also involved in the control of breathing, pulse, erections, and other involuntary responses, it directs feelings of fear, pain, and pleasure. The fronto-insular cortex is active when our brains empathize with the suffering of others, as when a baby cries or another person is hurt. It also fires into action when we try to spot attempts at deception, to help us detect when someone is lying.

In whales, spindle cells are also found in the frontopolar cortex, but no one knows what they are doing there. Whales have come to have these cells on a completely different evolutionary pathway from humans. Such a result of natural selection is called "evolutionary convergence," as the same end has been achieved on a radically different path. Whales have had these cells for at least twice as long as we have had them, and they may have three times as many of them in proportion to the mass of their brains.

"Spindle neurons transmit highly processed information," says Hof. They help us remember emotions and insights, and enable us to care for others beyond our immediate needs. "It's tantalizing to think that the songs and vocalizations of whales might be based on similar processing."

How does this jibe with the lack of information that our engineers found? The presence of spindle cells suggests the whales may be adept at communicating *emotional* content to one another. And that's nothing that can be translated into bits per second, minute, or even hour. It's like a ton of bricks hitting you in a single moment. Something like that could easily be contained in such amazing sounds. We opened up their brains and didn't find the reason why whales sing, but there is evidence that they may be able to *care* about our question.

It's a tedious job, but someone has to do it. Get up early each day, head down to the water's edge. Check the engine, review yesterday's notes. Start the engine, steer the boat carefully out of the crowded Lahaina Harbor, and look for humpback whales—singers in particular, lone males suspended fifty feet underwater performing their thirty-minute songs. Jim Darling has been out on the water listening to singing humpbacks for nearly twenty seasons, longer than anyone else. Headphones in place, he gazes out on the horizon with a searching look, aboard a boat whose name offers up his scientific creed: *Never Satisfied.*

There is always another question that should be asked and a better way to answer it than we've yet come up with: Why is this song so deep and beautiful? What is it trying to say? Who is listening?

Let's review the various theories that have been put forth to explain humpback song, and then explain why Jim Darling doesn't buy any of them. Once it was determined in the late seventies that only the male whales were singing, early researchers assumed the song was a sexual display to attract females. Only problem was, no one ever saw a female

*Fig. 35. Jim Darling aboard the* Never Satisfied.

approach or take interest in a singing male. So other theories were proposed: perhaps the song is a navigational aid, helping whales find their way across thousands of miles of great oceans. Perhaps the very presence of the sound helps space the males the right distance apart from one another (since they usually sing when alone). Maybe it's something purely physiological that helps to synchronize the estrus of females. And, because of the range of sounds and their rhythmic qualities, each part of the sound might have a different function, especially a kind of sonar that helps the males locate females—the Mercado hypothesis.

Once again, there is little evidence for any of these theories. They are nearly as far-fetched as the explanation for whale songs in Christopher Moore's novel *Fluke*, in which the whales are giant underwater submarines commanded by a fiery woman who's the granddaughter of Amelia Earhart. Or remember the movie *Star Trek IV*, when Earth is under siege by hostile aliens in the twenty-third century? The aliens can only speak in the language of humpback whales, which have unfortunately gone extinct. So Kirk, Spock, Chekov, and company travel back to twentieth-century Earth, where they get ahold of some humpbacks from a San Francisco oceanarium and transport them to the future where their songs are able to save our planet for eternity. (I remember at the time the film came out, Paul Winter told me that a media blitz as big as this was sure to finally save these whales. He even made a record with Leonard Nimoy and Roger Payne intoning great whale quotations above the Consort's placid and solemn tunes.)

Darling has spent too much time out in the field watching whales to consider theories that just don't fit what he's been seeing. Females never approach a singing male. Males sometimes display competitive behavior over females, but never when song is sung. Time and again he saw other males approach a singing male, but they did not start singing in any kind of signal-jamming fashion, the way nightingales do, which would suggest territorial competition. No, the two whales would often go off silently together. Then they would separate. Song tended to temporarily gather males. Then they would resume their solitary ways and begin to sing again.

So the males seem to be singing for each other, and not against each other, a phenomenon tough for biology to explain, or even admit. The elaborate series of phrases and patterns, with all their rules and form, do have some social purpose, but something very unusual in the annals of animal communities—a music of cooperation.

In autumn 2006 Darling published a groundbreaking paper in humpback-song research, the result of many years of direct observation of the whales on their breeding grounds in the quiet waters around Maui, Lanai, and Molokai. Whereas previous studies tended to analyze the specific structure and changing patterns in humpback song, Darling took detailed field notes of what actually happens when one whale comes upon another who is singing, and he carefully told the stories of 167 singing whale/other whale encounters.

Eighty-nine percent of the interactions were cases like these, where a non-singing male would join a singer for a while, then go off on his own:

**Example 1: March 5, 2000.**
*A singer was joined, split, then serially joined two separate singers, then began singing again.*
This male, singer B, over a four-hour period, interacted with three, possibly four, males, for five-six minutes each over a distance of 13 km. The singer was joined, stopped singing, then itself joined and split from two separate singers (D and E), then began singing again.

**Example 2: February 9, 2001.**
*Traveling male joins one singer, bypassed the next singer and joined a third singer.*
This non-singing male B joined a singer A, split up, passed by another singer C (at 300 m), then moved to join a third singer D, and split up again. The longest interaction lasted 3 minutes and the lone male traveled over 6 km in two hours.

**Example 3: March 22, 2002.**
*Traveling singer stopped singing, approached another singer, split, then moved to join a third singer with which it paired.*
Whale A, over a total of 4 hours and 8 km, traveled while singing, then breached and stopped singing, passed singer D (within 300 m) then swam to join singer E with which it formed a pair. The distance between these individuals steadily widened as they swam in a generally parallel course.

Eleven percent of the interactions were of the next type, where first a non-singing male (now called a "joiner") would join the singer, and together they would approach a female with her calf, and then escort them:

**Example 1: March 26, 2000.**

*Joiner plus singer formed a pair, then the pair joined a singer (escort) with a female (with calf).*

Lone whale A moved straight to singer B, which stopped singing as it was joined and the pair (A and B) moved toward and joined singer D, which was escorting a cow and calf while singing. There was a six-minute interaction under water. On surfacing whale A and whale B departed in different directions and left group D, the ex-singing escort with cow and calf, which moved off slowly in a third direction. . . . Less than one hour later D was singing again and an hour after that was joined by two other adults, which led to a surface active group, in which the interactions between individuals were not clear.

**Example 2: March 26, 2001.**

*Joiner and singer joined to form a pair which then joined female (with calf).*

Singer 2B was joined (joiner C) then the pair traveled immediately to a cow with calf (no escort) that was initially 400 m distant from the singer. Overall, this group D (cow/calf, ex-singer and joiner) acted like a surface active group in terms of speed, changes in direction, lunges, and bubble streams during the 60 minutes of observation. The position of the individual whales in the group was consistent, with the ex-singer chasing and ex-joiner apparently blocking the female.

**Example 3: March 29, 2001.**

*Singer stopped singing, joined another singer, split and joined escort with cow and calf.*

Traveling singer C stopped singing and joined singer B for less than 3 minutes before splitting up. Ex-singer C then raced 6.5 km (7.5 km per hour) until it joined a cow/calf and escort E. (A loud song was recorded in the area just before this joining, and was very likely the escort but this was not confirmed before it stopped.) Ex-singer C interacted with this group for just 3 minutes during which there was much disturbance and movement of the cow. On one surfacing ex-singer C was perpendicular and directly in front of the female apparently blocking her forward movement (rather than fighting with the escort). The male escort was extremely excited and chased the cow with its nose underneath the cow's tail, the cow moving very fast with a noticeable wave off of her fluke.

For the first time, we have direct reports of how non-singing whales react to singing ones. Sound does not happen without a context, as the information theorist would want us to consider it. The song leads to very specific behavior, which Darling, together with grad student Meagan Jones and underwater photography pioneer Flip Nicklin, are the first to rigorously document. Here's a summary of what they saw.

Males most often are the ones who approach another singing male, and when they meet, the singing stops. Most of the time the meetings were peaceful, but on the rare occasions when the singer was with a female before being approached, there was sometimes tension. The meetings were usually brief, then the whales would separate. Sometimes, a pair of males would go off together and be ready for the next encounter. Non-singing males tended to go from one singer to another over a range of several kilometers. If two singers were near each other, often one would stop singing and swim over to listen to the other for a short while. Singers would sometimes chase females, but the females would show no interest. Behavioral patterns did sometimes emerge: solo singing, then joining, then males happily traveling together, then one male leaving to join a group that included a female.

Darling and his team are onto something. Humpback whales do listen to each other, and they come and check out the singer. Those that do so are always other males. "Singing apparently broadcasts the specific location of the singer and, most certainly, other adult males use that information." Even though the song's bit rate may be low according to the engineers, the whales are finding something valuable in it.

How might the song organize the males? Darling asks us to consider the two unusual qualities in the sound and how the whales behave: (1) The song changes during the breeding season, and (2) there are continual associations between males that are not aggressive. So there is some kind of "drive for novelty" within these male whales, and there is some suggestion that they may cooperate.

Darling and his team speculate that "the song dynamic provides a real time record of the relationship between males." Not in terms of spacing them correctly across the ocean, but because when they are together, all sing the same song. "It works as a type of feedback loop, in this case based on emulation and incorporation of song change: *the more closely associated the males, the more similar the song; and the more similar the song the more associated the males.*" It's like a secret code in high school: if you're with us, you know the password. This way of getting with the program requires nothing more than taking your song to the next level.

He doesn't know if the whales use this quality to distinguish neighbors from strangers. Or if the whales decide some new songs are better to adopt than others. The newness itself may be what matters, as Noad's Australian case study implies. At least it's a simple enough explanation, where no one needs to be in charge. The mutual recognition process works as long as the song keeps evolving from week to week, year to year, and as long as any changes are accepted by all the males present.

What about the other 11 percent of singer/joiner interactions, where the two males, the performer and his audience, would then join a female and her calf, and she would tolerate them both? Darling doesn't want to ignore this situation; it's statistically significant enough. Perhaps the two males, now both knowing the right song, are now escorts for the mother and daughter to fend off rival males who might not yet know this season's tune. Or maybe mating itself requires some kind of cooperative behavior among males. We don't know; we've never seen it happen.

Cerchio, one of the peer reviewers for Darling's paper, feels that these instances when males approach a female are much more important than Jim gives them credit for: "This represents the largest sample of observations of *females* associating with males. Jim de-emphasizes interactions between males and females so he can emphasize male-male interactions. He chooses to ignore it, and then what he sees as non-aggressive behavior gets translated into cooperative behavior, which he admits is also something that's hardly ever been observed with humpbacks! You get on a pair, you know they are males, we follow them, and they approach a female. When they get there, these would-be buddies start to compete!"

Cerchio still believes the mating system of humpbacks is like a giant lek, where the males are all singing together in a vast chorus, spread out over an area of several hundred square miles. "We humans are tiny creatures in the middle of it, like an ant in the middle of a fallow deer lek, we're unable to perceive the whole thing." We're just imagining there are single whales singing alone. If you randomly pay attention to one animal who's not getting any action, you're going to miss the forest for the trees, or the ocean for the whales, so to speak. Cerchio remains confident in the mainstream theories of mammal mating strategies: "We have very strong evidence that there are competitive interactions between males. You can't really discard that. There are a *few* indications of cooperative behavior, but I haven't seen any in my own twenty years of studying these animals."

Peter Tyack spent several years out in the field with Darling in the 1970s, but he doesn't want to give in to his friend's new evidence. "Jim's quantitative descriptions are correct, but he just has a naïve observation-bound perspective. The better way to do science is have a theory and hypothesis, and then make predictions about what you are likely or not likely to see. In most species males interact with each other far more often than with females, but that doesn't mean the less frequent interactions aren't of great importance."

Yet Darling holds his ground. "Everyone else keeps promoting the old sexual selection theory, for which there is hardly any evidence. I wonder if they keep on doing it just to get me mad." He is also a cautious observer with a theory guiding what he sees. "It's not like I really *believe* what I've come up with," said Darling to me over a cell phone from Tofino in the middle of a hailstorm, "but the alternatives have no evidence at all to back them up. We're way ahead of everyone else on that score, because we've been watching what these whales do when they sing for years."

No one approaches the song of the humpback whale with innocent ears. A biologist with sexual selection on the brain will want to hear a chorus of competing males, their overlapping song patterns rumbling down into a vast undersea choir in a sexual fray. A true behaviorist will want to follow a singer closely and watch what he does, focusing more on what can be seen than what can be heard. And a musician will hope to get one singer alone, to best appreciate the full beauty and form of his music. Then I will try to find a humble way to join in.

# 7

## MOBY CLICK

### Sperm Whales Got Rhythm

The single most familiar image of a sperm whale is still the fierce, ghostly persona of Moby Dick, that albino beast hell-bent on human destruction. The whaling industry prized the sperm whale for its vast quantities of oil, which were used to light lamps, to lubricate machinery, and to fashion into perfumes and potions. Our civilization was fueled by whales for a hundred years, but the animals were never easy to find. The sperms were wary and knew how to elude us. Already in 1835 Thomas Beale, in the first book ever written on the animal, knew that they must have some kind of sophisticated means of communication, "by which they become apprised of the approach of danger. . . . The distance may be very considerable between them, sometimes amounting to four, five or seven miles. The mode by which this is effected remains a curious secret." No human could hear it.

By the early twentieth century, whaling got more efficient and ruthless, but whale oil was no longer the necessary commodity it was in the early nineteenth century. Petroleum had long supplanted spermaceti as the main source of oil to lubricate our progressing world. Twentieth-century whaling, with improved technology, was much more interested in the giant baleens: blues, fins, sei, and humpbacks. Sperm whales don't taste nearly as good, so we lost interest.

Until scientists started studying their social lives. These great beasts do not sing, but click—quick, rhythmic taps, some quiet, some the loudest sounds we've ever known an animal to make. *Click. Click. Click. Clicccccreeeeaakkkk. Click. Clickety click* [pause] *click.* Subtle, but exact. The study of these ticking beasts in the world's deepest waters has led to the greatest question in our journey to understand whales: Do whales have culture? More specifically, do they have ways of life that are taught and learned from one generation to another, where distinct populations behave differently in ways genes cannot predict?

The head of the sperm whale is a huge melon filled with spermaceti, the source of oil prized by the whaling industry. Although the early whalers thought the stuff had something to do with sperm, it turns out this giant organ is mostly involved with making and processing rhythmic clicks. Of course female sperm whales, who make most of the sounds, have spermaceti too. (Why imagine semen in the head—too many men alone at sea too long perhaps?) Whalers called the top part of the head the "case"; scientists now call it the spermaceti organ. The hunters called the bottom part the "junk"—that name stuck. Both sections are soaked in the most valuable organic oil known to man. Barrels of the stuff are still used to grease the workings of the most precise machinery.

The sperm whale is certainly an animal of extremes. It's the largest predator on Earth, and the deepest diver of all mammals. There is great sexual dimorphism: only males have teeth, and they can be 30 percent larger than females, up to fifty feet in length, weighing forty-five tons. Females and calves live in the social groups that have received the most study; males spend much of their time hunting alone. Sperm whales have the largest brains known on the planet, and they are known to do battle with one of the largest phantoms of the sea, an enemy only once captured alive—the giant squid.

Whale culture has been hypothesized in orcas from their clan-based whistles, but clicks in all cetaceans are generally thought to be used for echolocation. Sperm whales use their clicks to locate their prey, which is mainly squid if they are available, but they also eat many kinds of fish in areas where squid are scarce. The whales also make other kinds of clicks, unevenly patterned, soft Morse code-like ticks. Until recently they were a complete mystery, but now we can use these sounds to tell which clan a whale is from. They may also be part of some vast, underwater rhythmic composition.

Like most of what goes on inside large whales, we don't know how the spermaceti exactly produces clicks and creaks, but it is believed that air is forced rapidly through various passages inside the whale's head and becomes focused by a series of "acoustic lenses" present in the junk. Some of this information has come from experiments in which sounds were played into the head of recently expired whales! During a lifetime a single sperm whale may produce more than half a billion clicks.

Although whalers had long reported that sperm whales travel in groups and that they exhibit interesting group behaviors, it was generally thought that it would be too arduous and expensive to follow the sperms on long-range studies because they are so widely dispersed in the deep sea. But in 1981 the World Wildlife Fund decided to support the research of Jonathan Gordon and Hal Whitehead, who started tracking and identifying individual whales off of Sri Lanka on a thirty-three foot sloop called *Tulip*, a craft almost the exact length of the average female sperm whale. It was the first step on what has become a career observing sperm whales thousands of miles offshore. Over the last forty years Whitehead has taken several more epic ocean journeys, some with his wife and children, with the sole purpose of recording sperm whale behavior, especially their wide range of sounds.

Whitehead is a tall, lanky character with bushy reddish hair and beard, who looks ready for the next adventure. His book *Sperm Whales: Social Evolution in the Ocean* is the best work I've ever read that focused on a single animal species. It is scientific in tone, comprehensive and technical, but it makes for fascinating reading. "I'm touched by your admiration," says Whitehead. "All I've been trying to do is getting to know the whale itself. That's the fundamental role of science in all this, and there is so much we don't yet know."

The 30 coda types with an approximate representation of the modal pattern for each type (except "Var" coda types for which an example is shown), the number of codas analyzed, and their classification into 4 classes (short, long, regular and plus-one)

| Name | Description | No. | Sh. | Long | Reg. | +1 |
|---|---|---|---|---|---|---|
| "3R" | | 625 | S | | R | |
| "3a" | | 131 | S | | | |
| "3b" | | 272 | S | | | |
| "1+2" | | 90 | S | | | |
| "2+1" | | 20 | S | | | P |
| "4R" | | 209 | S | | R | |
| "3+1" | | 51 | S | | | P |
| "3++1" | | 68 | S | | | |
| "4L" | | 69 | S | | | |
| "4Var" | | 72 | S | | | |
| "4+1" | | 227 | | | | P |
| "4++1" | | 77 | | | | |
| "5R" | | 269 | | | R | |
| "2+1+1+1" | | 53 | | | | |
| "5Var" | | 55 | | | | |
| "5+1" | | 160 | | | | P |
| "4+1++++1" | | 99 | | | | |
| "6R" | | 149 | | | R | |
| "6Var" | | 98 | | | | |
| "7R" | | 131 | | L | R | |
| "5+1++1" | | 56 | | L | | |
| "6+1" | | 61 | | L | | P |
| "7Var" | | 150 | | L | | |
| "8R" | | 109 | | L | R | |
| "8L" | | 54 | | L | | |
| "8Var" | | 98 | | L | | |
| "9" | | 79 | | L | | |
| "10" | | 63 | | L | | |
| "11" | | 31 | | L | | |
| "12" | | 18 | | L | | |

*Fig. 36. Weilgart and Whitehead's thirty-three sperm whale coda types.*

Sperm whales make two kinds of clicks: regularly spaced, even ones, whose function is generally believed to be echolocation; and quieter, uneven rhythms, which were originally hypothesized to be signature sounds that identified individual whales. These are called "codas" (even though they don't always come at the end of something). On the first expedition aboard the *Tulip*, Whitehead and Gordon did not pay too much attention to coda clicks, since they happen only when the whales are socializing, and that is only about 20 percent of the time. That study ended in the mid-eighties because of the Sri Lankan civil war. Gordon went into conservation, and Whitehead married a whale-obsessed woman named Lindy Weilgart, who became his companion on a series of journeys across the Pacific Ocean, between the Galapagos Islands and the coast of Chile. The couple knew from old records that sperm whalers had done well there in the past.

As their research progressed, Weilgart and Whitehead were able to create a lexicon of sperm whale beats, analogous to the catalog of cries and shrieks assembled by John Ford for the Vancouver Island orcas. In the South Pacific they identified thirty-three different coda types, all little rhythmic riffs, from three regular beats, to two beats, a rest, and a third beat, to four beats, a rest, and a fifth, to seven irregular beats. The catalog of patterns makes quite a comprehensive chart, as shown in figure 30 on the preceding page.

From a musician's standpoint this looks like a list of possible rhythmic patterns, say the range of beats that an Indian tabla player might use, or a Latin percussionist playing timbales. The scientists, however, didn't know what to make of this information, since it seemed so anomalous. These clearly weren't echolocation clicks: they were neither extremely loud nor directionally focused in a way that might help locate prey. Nor could they be signature sounds, which would identify individual whales, since there are far too many whales, and far too small a number of different codas to reliably identify a single whale. "They're very strange," says Whitehead. "They have a clear and distinct pattern but they're very simple, at least as we hear them. It makes them easy to study, to quantify and to organize." So what were they for?

Weilgart and Whitehead sailed around the South Pacific, trying to discover how sperm whale behavior, and especially codas, varied over vast distances. "This was a lovely excuse to spend a year on a boat with our family," said Hal. For much of the trip, they saw no whales at all, even in areas where whalers always found them. They didn't expect that. The two researchers also imagined that the codas in the eastern part of the Pacific would be different from those in the west, due to genetic differences. But when they did hear whales, Whitehead and Weilgart heard something unprecedented.

There were groups of whales scattered all over the Pacific that seemed to be identifiable by virtue of the kinds of clicks they favored. Tissue sampling showed the whales in any one group shared mitochondrial DNA, suggesting they were related on their mother's side. But this is not enough genetic similarity to have inherited behavior. Instead, it suggests that sperm whale clicking behavior is indicative of a matrilineal sperm whale society—the whales seem to learn their group characteristic clicks from their mothers and their mothers' relatives. Each group consists of one, two, or three "units" of perhaps eleven animals each. The whole group includes up to thirty-five animals in total.

While on the water, Hal tends to collect the data and not analyze anything prematurely. "There are people who are very good at understanding what animals are doing in the wild, but I'm not one of them. I go out and collect the data, then take it all home. I can only pick up this stuff through analysis much later." This explains why it can take up to a decade for these whale science papers to get into print.

When he got to the stage of charting and tabulating which units used exactly which codas, Whitehead found a surprising pattern of organization at a higher level: certain units seemed to share a definite repertoire of clicks. Whales separated by thousands of miles knew the same codas.

"When you listen to these sperm whale groups," notes Whitehead, "you find they are genetically unrelated, they are all female, but they have the same codas. This posed a dilemma that we did not fully understand. The groups are only together for a short time, and they're made up of two or more independent units of whales. So how could they all have the same sounds?" They must possess the ability to learn rhythms, the beats that you need to be part of the band.

"Does each whale listen to all the different clicks from different groups?"

"Of course, how could they not? The codas are made almost exclusively by the females. The females have this whole social system that they have to mediate. With other toothed whales like orcas and belugas you have a wider range of sounds, but the same division into sonic groups. But only sperm whales have this matrilineal society where the males are chucked out."

That's how Whitehead and Weilgart identified twenty units of sperm whales throughout the South Pacific, which they named "A" through "T." Different units would associate together to form temporary groups, and each unit could most easily be distinguished by the coda clicks they were making.

"As I tabulated the data I was just blown away!" It made more sense than Whitehead expected. "I found these two sets of units with very distinctive coda repertoires, which is something we had never noticed before." Groups of whales that shared the same mother made similar sounds. The kind of click favored by a group could identify its matrilineal family. Weilgart and Whitehead named this the "acoustic clan," similar to what Ford and Spong found with orcas, but with clicks, not cries. Figure 37 shows the kinship chart that shows how the initial twenty units fit into three basic clans, distinguishable by sound.

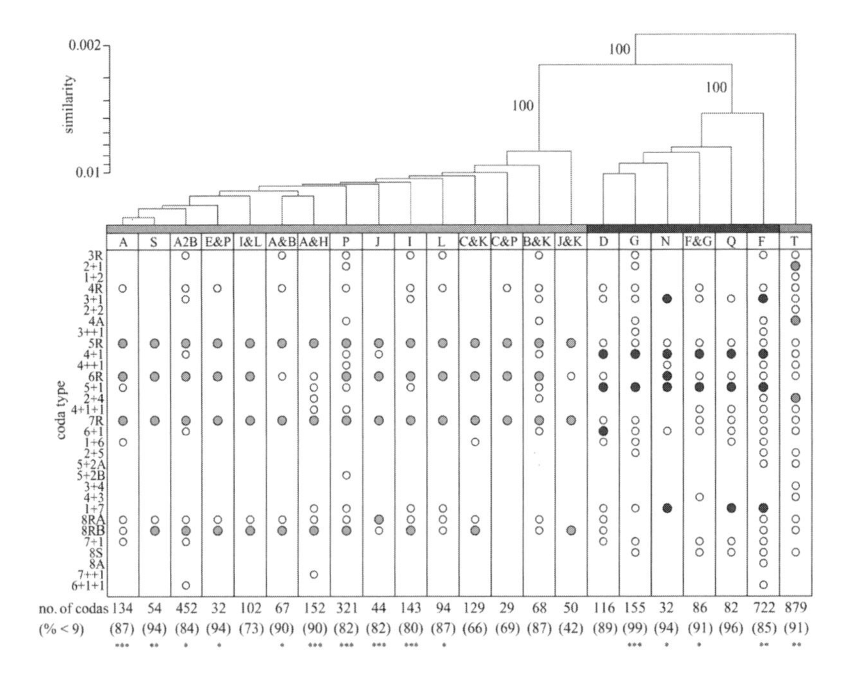


*Fig. 37. The coda types led the researchers to identify three acoustic clans*

Although it resembles a fingering chart for a flute, this diagram shows which particular patterns are favored by which units of whales. Some units are acoustically identical, but they were found separately. The filled-in dots reflect codas that make up more than 10 percent of that unit's repertoire.

The most basic analysis shows that the units on the left favor regular rhythmic codas, mostly 5 beats, 6 beats, 7, or 8. These were named the "regular clan." The next set of units favors a little unevenness in their phrases, especially 4+1, 5+1, and the syncopated 1+7. This was named the "+1" clan. The last unit doesn't fit in anywhere, but it turned out to relate to other units found thousands of miles away.

The clans span vast areas of deep ocean water, and in any one area two or more clans may overlap. Further analysis revealed that sperm whales tend to associate only with other units in their clan. Each clan has other distinguishing behaviors besides sound: the regular clan, found in the eastern tropical Pacific and also off of Chile and Ecuador, tended to travel tortuous paths and dive synchronously. The +1 clan, found mostly around the Galapagos, tends to swim more directly, and they do not synchronize their diving. Eventually Weilgart and Whitehead were able to identify four distinct acoustic clans by combing through all their South Pacific data.

What are the implications of all this coded behavior? That sperm whales, who learn these clicks at the level of the clan, group, and unit, possess a form of culture. Whitehead believes he can explain why the sperm whale has so large a brain: "I don't think it's for making clicks," he grins. "The case and junk are big enough for that. But the brain may be for processing clicks, because it has to detect the click out of a lot of noise. As a percentage of body weight the sperm whale brain is not terribly big—the bottlenose dolphin is second only to humans on that statistic."

He makes an even wider speculation: "Perhaps the sperm's brain must be large to deal with social and cultural information. They live in complex societies. A lot of what they do depends on learning from each other. Social learning is something that we as humans do all the time. Some of us are a lot better than others, and I suspect the same thing happens in sperm whales. There's an enormous selection process to do well, so maybe that's how the brain got so big."

"Do the clicks themselves convey any particular information?"

"I think they do. It may be as simple as 'we're best buddies,' but it may be something more. The mother of the calf has a somewhat different coda, maybe they have to do with raising the offspring. We don't really know. The coda is a reaffirmation that we are from the such-and-such a clan. But why would they need more than one coda for that? This is what we need to work on next."

Whitehead first published his hypothesis in the journal *Behavioral and Brain Sciences* in 2001, with an article written together with his student Luke Rendell. The notion that sperm whales could have different cultures, based on a barely perceptible difference in the tiny rhythmic clicks of separate groups, sent a revolutionary challenge to the world of animal behavior studies. Differences scientists could hardly perceive in the field were being put forth as evidence of learned, social behavior in a huge marine mammal that we had previously only taken seriously as a source of high-grade oil. Whitehead and Rendell suggested that those who study marine mammals might consider their work more like ethnography than zoology, where each population studied is a distinct society with its own unique, defining behaviors—like a tribe.

The argument for culture in cetaceans is quite simple. Culture is basically defined as a set of behaviors that are learned by and maintained by a group of animals, independent of or not determined by genetic inheritance. If a pod or clan of whales learns certain ways of living from the others with whom it associates, and maintains those qualities independent of environmental factors, then those attributes are culturally determined.

Some groups of killer whales hunt mammals in packs, others more calmly eat fish. They live in the same areas, but with different lifestyles. They are the same species, but they have learned and more importantly *maintained* these differences over generations, even though they interact with and acknowledge the other populations of whales and their ways of life. Pacific humpbacks sing their songs while suspended fifty feet under water, while humpbacks in the South Atlantic, as observed off of Brazil, stick their tails in the air and wave them gently back and forth as they perform their great serenades (makes them a lot easier for researchers to find).

Cetaceans can imitate the behavior of other animals. A wild dolphin can hear another dolphin's signature whistle and copy it right back. Individuals of many whale species can pick up new behaviors and then teach them to other members of their pods. A new humpback feeding behavior was observed off Maine, and scientists were amazed to see it picked up by all the whales in a matter of weeks.

The article was extraordinary enough to be published with thirty-nine separate commentaries and critiques. Anthropologist Jerome Barkow went so far as to say that cetacean and animal culture, with its rather simple learned behaviors maintained by a social group, could be called "normal" culture, while human culture, with its vast array of learned traits, is really a kind of "hyperculture" unmatched in the natural world. Evolutionary psychologist Robin Dunbar disagreed. He doubted that cetaceans can share meaning in a cognitive way, so he doesn't think "culture" is the right word to describe how they pass behavior on from one generation to the next.

More supportive comments tended to suggest that there is much more culture in the animal world than Rendell and Whitehead realized. Parrot expert Irene Pepperberg noted that her birds (especially the late, great African grey parrot Alex) certainly possessed social learning ability, in the wild and in captivity. The astute bird scientist Peter Slater pointed out that the question is not whether or not whales have culture, but *why* certain species have learned social behavior and others have not, since natural selection doesn't seem to adequately determine this quality. Primatologists David Premack and Marc Hauser argued that we need to figure out which differences in the behavior of different groups of whales are trivial and which are not. What do the differences in humpback song and sperm whale clicks actually mean? That's what we should focus on, to separate arbitrary variation from real differences in culture, like the orcas' group choice to hunt either mammals or fish.

If cetaceans really possess culture, then perhaps human cultural approaches might actually help us understand them. Sperm whales certainly click, often in regular rhythms, and sometimes in irregular rhythms. Individual differences in their sounds have been mapped out by Whitehead, Rendell, and Weilgart. But most of the time these clicks overlap upon each other, forming a confusing jumble of rhythms, difficult for us to make sense of. Difficult for scientists, at least. Not so tricky for someone skilled at following a jumble of beats going on at the same time. Who might that be?

While Whitehead and his team were reporting the amazing range of sperm whale rhythms, French scientist Michel André, working in the Canary Islands, had some questions. André, while listening to sperms clicking throughout the day, noted that codas form a very small portion of their repertoire. Most of the clicks are regular beats, what Whitehead and most researchers tended to call echolocation clicks.

But surely, thought André, the whales would need to communicate to each other throughout the day, not just occasionally. The regular clicks, he surmised, ought to have a communicative function as well. These more prevalent regular beats might be more than echolocation clicks. They are regular beats at differing tempos, overlapping rhythms, all at different speeds, fitting together. Listening, he could make no sense of it. Mathematically analyzed, it was still too confusing.

André remembered that European musicologists, when first visiting Africa, could not understand how a single musician in the midst of large groups of drummers in countries like Senegal could keep their own part going in the presence of so many other contradictory beats. In fact they've been maintaining their individual signature rhythms in the din of the crowd since childhood. From years of practice, each drummer knows how his pattern sings out in the spaces between all the other patterns. One must be an expert in the discernment of rhythms to successfully play this music.

With this in mind, André invited Senegalese drummer Arona N'Diaye Rose, one of the many sons of the great Doudou Rose who intuitively grasped the signature whistles of dolphins a few years before with André's colleague Cees Kamminga in the Netherlands. Arona Rose listened to a recording of a four-member unit of vocalizing sperm whales. The *sabar* master was immediately able to distinguish the beat of each of the four clicking whales from the others. He also believed that what the scientists heard as cacophony was actually an organized rhythm, based on a dominant beat coming from one of the whales, which Rose felt was analogous to the signature rhythms marking the social structure of an African tribe.

"I couldn't believe it," said André to me at a café in Barcelona, close to his laboratory. "We knew there were four whales because we took notes during the recording, but all we heard was a confusion of clicks. I asked Arona how he could tell there were four different animals. He said, 'I don't know how, but I know.'"

When I first heard this story I had to smile—it is just the kind of wondrous tale that makes me hope that music and biology might work together. I passed it on to my friend Andrew Revkin, science reporter for the *New York Times*, and pretty soon it appeared in many newspapers around the world. Nature needs culture. Science needs art.

Animals spend a large amount of time making calls back and forth to each other, in pairs, groups, or all at once, in the striking dawn chorus of birds or a lek of chorusing frogs. As common as such behavior is, it has hardly been studied. We don't have the tools to accurately analyze the relationships between many simultaneous noises, and we tend to focus on the sounds, not the spaces between the sounds. But that's what rhythm is all about: silence, and the marking of pulses in the wake of the silence.

Musicians whose ears and minds are tuned to hear layer upon layer of overlapping beats as music, not noise, may be well trained to pick out the order behind the overlapping mess of sperm whale codas. Whitehead took the first step and identified the different kinds of codas. Now André and Rose want to study the relationship between each pattern, not with machines, but through an education in polyrhythmic listening.

Since that session with Rose ten years ago, André has been seeking funding to continue this work. But it's the same story many scientists have told me: it is always difficult to get support for descriptive work. Applied science, especially work toward managing whale "stocks" or populations, is always the easiest to fund. Something that combines biology and music, for all its intercultural promise, is harder to support. Will it be funded by research agencies or cultural exchange groups? Neither wants to touch anything so clearly on the charged border between one approach and its opposite.

Some scientists are suspicious of this news from the Canary Islands that has been heard 'round the world. Anything smacking of music is too subjective to be of much use to science. "Michel André?" Hal Whitehead shakes his head. "I have no idea what he's talking about. But Luke says he understands him."

I asked Luke Rendell what he thought of André's work, and he was far from enthusiastic: "André analyzed just eighty-five seconds of data, producing periodograms of click intervals from just three whales. That's about a hundred clicks from each whale, from an average per dive production of two thousand. . . . You simply can't characterize a communication system as complex as the sperm whale's with so little data." Plus, nearly all sperm whale researchers are convinced that the "normal" sperm whale clicks are primarily for echolocation purposes, and not for the cultural rhythmic activity Rose hears in them. In fact, this paper, for all the inspiration it has given those like myself and the world media, has only been cited twice within the sperm whale research community. Those in the know have not been impressed.

But André will not give up. Look, he points out, in the Canary Islands, we rarely heard codas when the whales were resting and socializing near the surface. The echolocation clicks were heard constantly, so perhaps they have a social role as well. "The codas, until now defined as the principal acoustic support of communication in sperm whales, actually seem to be only a part of a whole system articulated by the continuous emission of usual click trains, which are the sounds usually made by the whales as they socialize. The famous codas are barely heard!" As to why he only looked at eighty-five seconds of clicking sounds, André says there was enough complexity in even such a small sample to illustrate his point, but the critique is fair. "We're currently working on these very issues, and the next paper is in the works. I am convinced rhythm is very important to these animals, even in the codas."

André has continued to publish on the rhythmic structure of codas, suggesting that there is much more variation in pattern than the Canadian team would admit. He is concerned that Whitehead would report two click trains as being "3+1," three regular and one irregular, even if the interval between the three regular clicks was different in each case, say, one triplet faster than the next. "I've talked to Rendell and Whitehead about this, and they realize there is a basic flaw in their analysis." André has been trying to come up with a more mathematically rigorous way to categorize clicks, as a prelude to being able to study the relationship between clicks coming from different whales with greater accuracy.

It would also be a method that takes account of the precise rhythm going on between the clicks, how they fit into a larger patterned context. This is what Rose heard happening, and André feels we should not ignore it. "We need to study the whales' perception, not our own perception. Scientists are more used to counting, so we count. We have to learn from the insights of African drumming to perceive the value of rhythm at work in the clicks themselves." So at the same time as trying wild ideas like working with drummers, he is trying to develop better and more precise mathematical tools.

It is easy to choose equations that show you what you most want to see. Whitehead admits his team has been so busy trying to identify the individual coda patterns that they have not spent much time studying the relationship *between* codas. In his wife Lindy Weilgart's thesis, there is brief mention of the occasional phenomenon of "echo-codas," where one whale clicks a pattern, and another copies it exactly, a tenth of a second later. But this was never investigated in greater detail, and they reported it only tentatively: "The ordering of codas within exchanges is to some extent non-random, suggesting conversations, but we do not know what information is being transmitted."

André wants to spend more time analyzing the spaces between sperm whale clicks. It's going to take musicians, scientists, and whales spending a lot of time together. "Sure, it's subjective if a drummer just listens once, but if I ever get to work with Rose, for several months' time, learning his perception, and his approach, to analyze the combinations, then I hope to learn something from his rhythmic intelligence that has been passed down through many generations. Yet we still don't have the funding to bring a drum master onto our team. Rose was certainly on to something, he immediately *sensed* a musical sense of culture that Whitehead and Rendell had to spend years in the laboratory to uncover."

The very idea that sperm whale tapping makes sense to both musicians and scientists should clearly signal its importance as a code worth deciphering. At the moment, only about ten people in the world have the slightest understanding of these giant beasts' code-like tappings. How to figure them out? The problem with whale science is the same problem Brian Eno pointed out ten years ago about the world of digital communication: "It's got to have more Africa in it." This is a perfect opportunity for a second generation of musical scientists, and scientific musicians, to delve once more into the mystery of whale sound. We know both these methods can help us cross the border into the cetacean world.

In July 2007 I attended the Third International Workshop on the Detection and Classification of Marine Mammals Using Passive Acoustics, put on by the United States Office of Naval Research and the New England Aquarium. In preparation for this gathering, signal processing specialists, many of them physicists and engineers, not biologists, received a data set of clicking sperm and beaked whales that they proceeded to decipher, using algorithms of their own invention. Who would turn out to be best at separating one clicking whale from another, and tracking their paths through the sea? Would mathematics be more accurate than human observation?

Paul Baggenstoss of the Naval Undersea Warfare Center tried to separate superimposed click trains, "seeking to minimize an aggressive criterion of optimality." He computed the distance between every pair of adjacent clicks and looked for numeric anomalies. George Ioup, from the University of New Orleans, discovered that there is a common spectral pattern to each coda of sperm whale clicks, making it possible to identify them automatically using algorithms called self-organizing maps, which can scan the data set to figure out which whale is producing which coda, and thereby aid in estimating the total number of whales.

When I mentioned to these two researchers the possibility of using similar approaches on African drumming, and the notion that drum experts might be able to detect click patterns with their heightened sense of rhythmic attention, both scientists were extremely interested, much more so than most biologists I had met. Signal processing analysis is much more like a game or a puzzle, and there is a sense of tinkering and playfulness to the whole field, in military and academic worlds alike. So perhaps the *Fourth* International Marine Mammal Detection Conference will invite some Senegalese drummers. If they invite clarinetists who jam with whales, it's going to be active, not passive, acoustics.

Believe it or not, there are a few examples on record of humans and cetaceans cooperating to form an interspecies sense of culture. In the town of Laguna near the southern tip of Brazil, fishermen have developed a unique method of fishing for mullet that works only with the help of wild dolphins. The water is especially turbid, and it is impossible to see the fish. About forty fishermen head out each day, with one to four dolphins assisting them. The men position themselves in a straight line parallel to the shore, each one holding a circular nylon throw net. The dolphins swim several meters farther into the ocean than the fishermen, moving back and forth on the surface. A dolphin suddenly submerges, diving seaward, and the animal comes up in a few seconds, heading back toward the line of fishermen. Just out of net range he dives deep down, and the fish he's been chasing head straight into the nets. "The men rarely cast without a dolphin's cue, and the fishing method seems to be initiated by the dolphins," say Karen Pryor, Scott Lindbergh, and Raquel Milano, who wrote this up in *Marine Mammal Science* in 1990. After the men have caught what they need, the dolphins eat the addled fish that remain. This behavior, passed down from generation to generation of people and dolphins, has been going on since 1847.

An even more remarkable story comes from the town of Eden on Twofold Bay by Australia's Pacific coast. For more than a hundred years the Davidson family hunted humpback whales from small rowboats with the help of a pod of killer whales, who learned to alert the whalers to the presence of large baleens by flopping their tails rapidly on the waves. The orcas would work together to round up the confused prey, aiding the humans like a second whaling boat would. Together orca and human would make the kill. The orcas, known as the "Killers of Eden," would be rewarded by getting to eat the tongue and the lips of the humpbacks. The rest was left to the whalers, who used the blubber to make soap, pencils, candles, and crayons. The leftover meat was turned into pet food. This arrangement was taught from generation to generation of humans and whales, all the way until the 1930s, when the last killer whale in the pod, Old Tom, died.

These tales demonstrate that whale culture can aid human culture in a most practical way. Today, whale voices continue to inspire us artistically. Jim Nollman, together with Sam Bower, the director of the environmental art web site www.greenmuseum.org, sent out samples of beluga, dolphin, and sperm whale sounds to about fifty electronic musicians. Seventeen of their pieces, from ten countries, were chosen to be released in 2006 on *Belly of the Whale*, a compilation from Important Records.

Kim Cascone, who once worked as sound editor for David Lynch's *Twin Peaks*, created a mix of rich ambient booms and alternating rhythmic clicks, evoking something far-off and swirling from the source material of whales. The British electronic composer Scanner, who also composed an anthem for the European Union, combined the restless patter of sperm whale clicks with his characteristic grave minor synthetic harmonies, overlaying beluga shrieks that turn into buckets of noise. Finnish electronica wizard Petri Kuljuntausta turned the sperm clicks into a rapidly pulsing loop that resembles the thrum of insects. Nollman himself added a carefully remixed version of his guitar twangs charming up orcas, perhaps his best track yet.

I was honored to be invited to play on this disk too. For my contribution, I decided to build a rhythm entirely out of sperm whale clicks, to provide a musical analogy to what I thought Michel André was talking about—a unique submerged tribal groove—upon which I then played the bass clarinet, trying to enter into the secret polyrhythmic channel of the whale. The version on the record is all instrumental, but the original includes spoken lyrics, intoning the whole story:

> In the year 2001, Atlantic Ocean, Canary Islands
> two scientists hear the clicks of singing sperm whales
> calling to each other down under the sea.
>
> Even with all their machines measuring the sounds
> and the times between the sounds
> no one can figure it out.
>
> So they call in an African master of drums
> He listens to these clicking whales
> Knows how to pick out one beat from the next
>
> He says, "I hear the beat
> There are many beats
> They are all beating 'round
> the sum of one rhythm."

And the master smiled and he said
"Well, there is a way to get inside the whale."
Inside the whale

Like George Orwell said in 1939
"Face it, you are inside the whale."
Stop fighting, or wishing, you control it.
Just accept it, endure it, record it.

How does it go? They ask
What do they say? They wonder
Where does it start? They murmur
How does it end? How does it end?

Inside. That's what it sounds like.
Inside the whale.

"No words on this disk," said Nollman. "These lyrics have to go." That guy is always giving me a hard time! Perhaps the less said the better when it comes to deciding what whale sounds mean. What if the sperm whales are making music themselves? This may be wishful thinking on my part. The reason for the rhythms of their clicks is still unknown, and might never be found.

In 2009 I released by own collective response to the diverse and beautiful sounds of cetaceans, sending samples of my and others' whale recordings out to a world of fine collaborators, for the project called *Whale Music Remixed*, available today wherever music is downloaded and streamed, and described in more detail in the liner notes chapter at the end of this book.

DJ Spooky is an internationally known DJ and conceptual artist. Markus Reuter plays in the Stick Men and the Crimson ProjeKCt. 3Corners of the World is David Rothenberg and Estonian guitar alchemist Robert Jürjendal. British sound artist Scanner has worked with Laurie Anderson, Radiohead, and the Bolshoi Ballet. Drummer Stephen Chopek has played with Leon Parker and Alana Davis. The White Sea Shamans convened once by the White Sea in Karelia. Gari Saarimaki lives in an abandoned schoolhouse on a Finnish island. The late Mira Calix is known for her operas composed out of live insect sounds. Lukas Ligeti's music is a unique fusion of acoustic and electronic, traditional and avantgarde. Cycle Hiccups hails from Petrozavodsk, which is a very nice place. Warren Burt is a major figure in the global avant-garde. Strings of Consciousness is producer Philip Petit. Francisco López is one of the most prolific soundscape artists ever. Ben Neill plays a mutantrumpet with two bells. Robert Rich is one of founding figures of dark ambient music.

Two pure whale recordings are included for listeners' own remixing pleasure, belugas from Russia's white sea, and one lone male humpback off the shore of Maui smack in the middle of mating season. A portion of the proceeds from this recording benefited the Whalesong Project, which used to broadcast humpback whale sounds live from Hawaii over the internet, as described in chapter nine.

Each click only makes sense in the presence of other clicks. They bounce from whale to squid, surface and bottom and back. There is rhythm in our breath, our movement, and the way the deepest questions repeat in our heads, over and over again. The whales may be wondering, too. The ocean reflects and absorbs their inscrutable rhythms.

# 8

# THOUSAND MILE SONG

## Moanin' in the Deep Sound Channel

Three-fourths of the Earth is covered by water. All the water on the planet occupies a volume of more than three hundred cubic miles, enough liquid to fill a cylinder two miles in diameter that would reach from here to the sun. It's no place for a human. In the majority of this wet, dense world, it is nearly impossible to see. But sound has amazing power there, especially the lowest, deepest tones of the largest whales, which may be able to travel from one end of an ocean to the other.

Leonardo da Vinci knew that if you were to place a long tube into the water with one end in the waves and the other at your ear, you would be able to hear ships traveling many miles away. That is the earliest European reference to the long-range power of sound in the ocean. As with many of Leonardo's insights, not much was done with this fact for many hundreds of years. In 1826 Swiss physicist Daniel Colladon stood at one end of Lake Geneva, and Charles Sturm, a French mathematician, stood at the other. At the time, underwater bells were being tested as a way to augment lighthouses to aid in bad weather navigation. Sturm struck a large bell under the water and flashed a light at exactly the same time. The interval between the beacon and the sound was measured at the other end of the lake by Colladon. Since light travels nearly instantaneously, the speed of underwater sound could be calculated. They came up with 1,435 meters per second—nearly five times faster than the speed sound travels in air. (The actual number varies greatly with salinity of water, and its temperature, but they were only three meters per second off from the speed of sound we today know existed in Lake Geneva at that time of year.)

How can sound move more quickly in a denser medium? Light and sound waves are very different animals: light waves are transverse, traveling at an extremely fast speed, vibrating in parallel to the direction they move, like ocean waves crashing against a beach, or rings of latitude on the globe. Light careens through open space, is slowed down by denser liquids, and stopped by solids; we can't see through a wall.

Sound waves, in contrast, are longitudinal, vibrating straight in the direction they're moving, like when you shake a Slinky and you watch the pulse move through the coils. When sound waves propel forward, they alternately compress and expand the molecules in the stuff they're passing through. Turns out that the denser the medium, the faster the molecules shake as the wave goes through it. (It's even faster in solids—put your ear to the ground and you can hear the buffalo stampeding miles away across the plains or a train far away on its track.)

Cetaceans and humans are after the same things when navigating the deep seas: we're both trying to find our way in a dark, lightless, foggy world. Although the ancient Phoenicians could figure out how far away a rocky shore was by ringing bells and listening for the echoes, only in the beginning of the twentieth century did we realize this approach might be more effective under water. It was the sinking of the *Titanic* that inspired us to develop a means to locate underwater objects when they could not be seen. The great ship might never have gone down if they had known that iceberg was coming.

The first successful "echo ranger" was patented in 1914 by Reginald Fessenden of the Submarine Signal Company. His contraption was based on an electric oscillator that sent out a very low-pitched tone, along with a receiver that listened for the echo. It could detect an iceberg two miles away, but it was an imperfect contraption, flawed in an essential way: it was incapable of determining in what direction the huge white deathtrap was hiding.

Underwater sound spreads evenly from its source, making it very difficult to find what we hear. Above the surface, our stereo ears can easily tell where a sound comes from. In water, sound seems to be present all around, like a total reverberating field. The simplest way to tell a noise's deep sea location is to move, and then listen to whether it gets louder or softer.

The early sound-pinging devices were no use tracking German U-boats in World War I, because the enemy vessels moved too fast. By the onset of World War II, all ships had depth-finders and better echo rangers that could bounce sounds off a hull or pick up the thrum of an enemy propeller from at least a mile away. The whole system was renamed sonar, meaning "*So*und *Na*vigation and *R*anging."

In the 1930s scientists began to measure water temperature and pressure and studied how their changing values affect the way submarine sound behaves. Athelstan Spilhaus at MIT invented a contraption called the bathythermograph, shaped like a small torpedo, which recorded both temperature and pressure changes as it was lowered through the water off the side of a ship. Researchers armed with this device learned that the ocean is divided into several distinct layers with different sonic properties. Closest to the surface is a layer warmed by the sun, whose temperature and thickness changes with the seasons, but is at a constant temperature throughout the layer. A sound moving through this layer goes at a reliably constant speed. Next comes the *thermocline*, a transitional layer where the temperature drops steadily with depth, and as it falls, the speed of sound slows down. Beyond a thousand meters down, the temperature does not vary much more. The greatest influence on speed in the depths is ever-increasing pressure, which causes the sound to travel faster and faster the deeper one gets.

At the end of the 1930s the military discovered that sound moving just between the surface layer and the thermocline tends to bend down toward the layer where the speed slows down. This refraction creates a shadow zone of sonic invisibility, a space where a submarine might be undetectable by enemy sonar. Bathythermographs were standard equipment on World War II submarines. The ships hid stealthily by adjusting their position depending on the changing thickness of the ocean layers.

Sounds produced close to the surface can travel many miles, but they eventually dissipate as the vibrating waves bounce against the surface and diffuse themselves into the surrounding noise. Sound produced thousands of meters down in the ocean might travel very fast, but the pressure is so great that noises rapidly evaporate into the thick liquid silence.

Six to twelve hundred meters down, temperature and the speed of sound are both at their lowest. In this special layer, certain very deep sounds can travel extremely far and still be audible. In the early 1940s, Maurice Ewing at Columbia proposed that low-frequency sound waves, under 100 Hz, which are less prone to scattering and absorption into the material that conducts them, should be able to travel extreme distances. He named this level of ocean depth the "deep sound channel." Very low-pitched sounds at this depth will reverberate clearly, never touching the surface or the bottom of the sea, and thus travel with little resistance across the oceans of the world. These low sounds become fixed in the deep sound channel and boom and thrum across entire oceans because they do not dissipate and do not reflect or refract off anything. The sound waves vibrate and travel endlessly, losing hardly any power or momentum as they go.

In 1943 Ewing exploded one pound of TNT and soon learned the sound of the blast was picked up clearly by hydrophones two thousand miles away on the coast of West Africa within the hour. Could the sounds of whales also travel this far?

After World War II, the U.S. Navy devoted considerable resources to the new Office of Naval Research. They wanted to track submarines from very far away. In the 1950s the Sound Surveillance System (SOSUS) was launched, a complex array of hydrophones fixed on the ocean bottom and connected to cables that went to secret listening stations set up on coasts all over the world. The Navy was able to hear a lot of things: what kind of submarines were out there, how many propellers they had, whether they were conventional or nuclear, and sometimes even the exact make and model number.

But they also heard a lot of other sounds that were of less interest to them. Deep booms, grunts, howls, squeals. Clicks, moans. Monotonously repeating super-low tones that didn't come from any machine they could find in their secret catalogs. What could be making them?

Eventually, the Navy realized they were hearing whales. They kept this knowledge classified for many years. As far as the Navy was concerned, these sounds were all just "biologicals," naturally occurring noises of no strategic import. Seamen were trained to identify them so they wouldn't get alarmed and think that a secret enemy sound was booming across the distant seas. No one outside the Pentagon got to listen to most of these recordings until decades later, when the Soviet Union suddenly collapsed and the Cold War ended.

Once scientists got ahold of these decades of recordings, they heard all sorts of things, which could be located with great precision. Just *how* precise they can be located is one of the aspects of the technology that remains classified information, but the system is surprisingly accurate for one that must operate in an underwater world known for its opacity. Geologists could locate underwater volcanoes and gain insight into how the ocean floor itself is constantly being created from molten lava pushing up from beneath the Earth's crust. Biologists finally had a way to track the movements of whales from the changing attributes of their thousand mile songs.

Chris Clark is a pioneer bioacoustic scientist, now at the K. Lisa Yang Center for Conservation Bioacoustics at Cornell University. In the bowels of the beautiful Laboratory of Ornithology, a celebration of birds and all things avian, there are a pack of scientists mostly studying whales, on the outskirts of Ithaca a few hundred miles from the nearest sea. In dark rooms they pore over piles of data, endlessly seeking similarities and differences that might bring meaning and purpose to the sounds they have collected over decades in the field.

Clark isn't exactly sure whose idea it was to open up the SOSUS data to the cetacean research community: "I never asked for it at all. It happened in the spring of 1991, during the first Bush administration. Al Gore, Sam Nunn, and Ted Kennedy pushed through a bill called the Dual Uses Initiative, to take military assets and try to use them for civilian science and environmental purposes. I don't know where they got that idea."

Out of the blue Clark received a call from a man at the Office of Naval Research named Dennis Conlon, who asked him to come down to Norfolk and take a look at their data. He hadn't had much experience with military culture before that. "I was amazed, it was like Dr. Strangelove, the secret war rooms, it was real! I got a message that we would have a special meeting to discuss the Dual Uses, and I saw my name was on the agenda and I suddenly saw I was on the program and had to give a talk about what I would do if I had access to this information. I hadn't prepared anything, so I got up and told stories."

And Clark knows how to tell stories. Like many scientists, he has more stories than he has ever had time to write up, stories of analysis but also speculation, ideas he has never had the chance to follow up on because he's too busy raising money to keep his lab afloat. When Clark speaks, he reveals a side of his experience that doesn't appear in his scientific writings. Here is a man who truly enjoys imagining what it must be like to be a whale:

"From a bowhead's point of view, migrating under the spring pack ice in the Arctic, vision is reduced down to several hundred feet, and hearing is everything." Bowhead whales are the only whales besides humpbacks who sing songs, a simpler song than the humpback but a song nonetheless. "It sounds as if someone is bowing a cello. *Hreeaph, hmmmmr. Hreeeaph, hmmmmr.* And you listen to this and you think that maybe you're hearing the ice grinding in the background, because they've incorporated these sounds of ice into their song! It's not that surprising, because the ice has forty different voices, it can sound like a freight train, like wolves howling, like babies crying. So there's this continuum of sounds from natural physical forces such as the ice growing and stretching, all the way to the animals who are traveling through this area, this very complex underwater world beneath the ice."

He explains his theory that the traveling whales gather themselves together solely through listening. "Communication by sound is the means by which your comrades—*excuse the term, admiral*—in front of you and behind and beside you, negotiate their way through the icefield. You'll hear *hmmmm* and then ten or twenty seconds later *mmmmmmh* and then a few minutes, then another, with space in between. We're all connected by sound."

Clark calls such a group of traveling whales an "acoustic herd," a group of animals that holds itself together with sound. Their music, like work songs or spirituals, keeps the culture going: "Imagine what it must be like to be a bowhead in twenty-four hours of darkness, working my way through an ice field where the folklore of my culture has told me that my grandfather has been trapped and nearly died in the process. This is not like migrating across the open sea, this is frozen ocean, under the ice, by bouncing their sounds off the ice, the whales may reconstruct an image of their underwater world."

He asked the audience to partake in a thought experiment: "Close your eyes and remember what's around you, and you create a picture based on your eyes. Now if I told you to close your eyes and tell me a picture based on what you hear, not on what you see, you'd find it very difficult. But that is what I think these animals are doing."

The Navy was impressed. A couple of weeks later, Clark was called down to Washington and named chief marine mammal scientist for the Dual Uses Initiative. "I told them, 'Wait a minute, you guys already know all this stuff. You produced cassettes to train Navy guys who distinguish a "biological" from a submarine, different guides for each ocean.' In every training manual that I've found for Navy technicians, every image I've seen, every spectrogram, in the background there were always whales!

They had terminology for things like the 'jezz monster,' when all the fin whale voices come to a crescendo during the summer months, and they just tried to block that stuff out. It was a pain in the ass to them, because it made the subs harder to spot."

On his first visit to SOSUS headquarters in 1992, Clark was ushered into a dark room the size of a gym with row after row of dot matrix printers spewing out scrolls covered with dashes and dots, old-fashioned representations of the sounds picked up by each of the hydrophones stationed all over the world's oceans. Clark peered to study one printout, and he saw a familiar blip near the bottom of the scale, "Exactly the right sound frequency for a blue whale. Then as I walked along the rows of machines, comparing the patterns from separate arrays miles apart on the ocean bottom, I noticed something else: they were detecting the same whale!" He felt a chill on the back of his neck as he realized that the Navy's system could be used to locate whales singing across an entire ocean: day by day, hour by hour.

Blue whales, the largest animals that have ever lived, have ten times as many neurons as we do, devoted entirely to picking up sounds below 100 Hz, way beneath the lowest notes of the piano. We can barely hear what they are doing. A blue will make one long, dark moan, lasting up to half a minute, and then wait exactly seventy seconds and make the same sound. Over and over again, in an exact but very slow rhythm, for days. In the Indian Ocean, they do it every 140 seconds.

Fin whales make a simpler sound, an extremely low pulse of 20 Hz repeated every thirty seconds or so, beneath the lower limit of human hearing. Because of the simple and regular nature of the fin whale's beat, it has been easiest to use in testing the theory of the thousand mile song. When this sound was first heard during the Cold War, some thought it was some secret frequency being used by the Russians to fill the oceans with standing waves that could allow the enemy to detect the position of our submarines. Ocean acoustics textbooks in the 1960s were still skeptical that such tones could be of animal origin. Suddenly the Navy started paying more attention to low regular pulses. Turned out these sounds were being made by the long, sleek Ferraris of the whale world, the fin whales. The sounds are so far apart from one another, we can only grasp their rhythm when they are sped up thirty times. Figure 38 shows a comparison of the tonal sounds *made* by some of the larger whales and by human beings. Click sounds reach much higher frequencies than these:

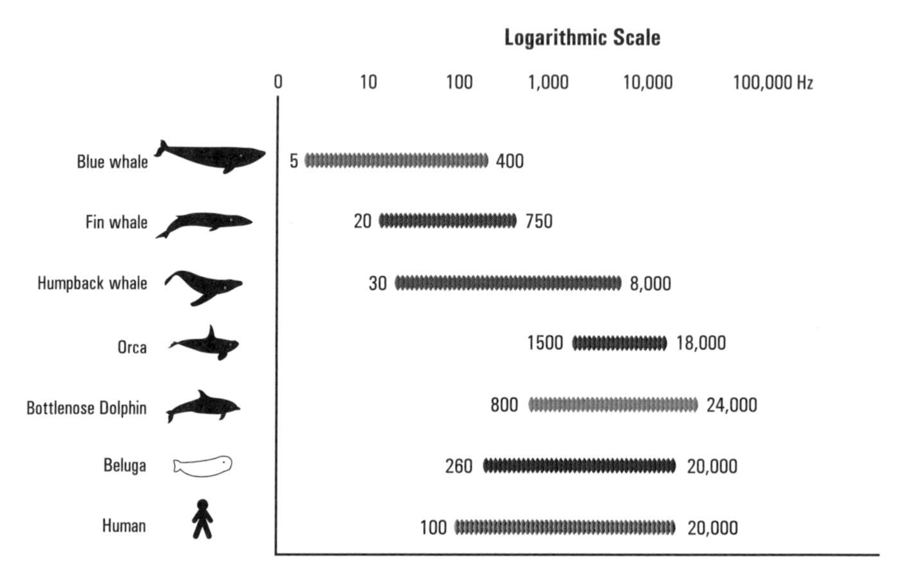


*Fig. 38. Comparative range of tonal sounds made by humans and various kinds of whales.*

And figure 39 shows how well whales and humans *hear* frequencies along the spectrum. The lower the threshold, the easier it is for each animal to hear that particular frequency:

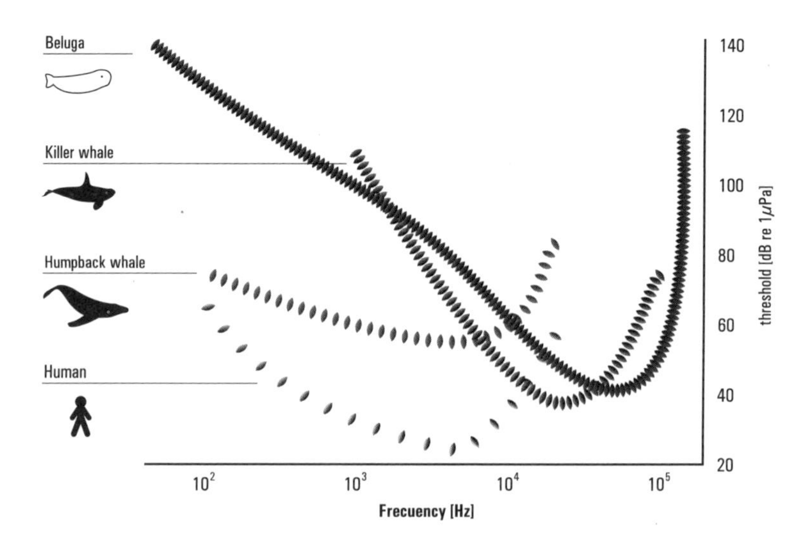


*Fig. 39. How well humans and whales hear different frequencies. The lower the threshold, the easier it is for the animal to hear.*

As early as the 1970s Roger Payne hypothesized that these largest whales who make the lowest sounds could conceivably communicate across entire oceans, because of the aforementioned properties of the deep sound channel, shown in figure 40 below.

Over the coming decades it was determined that only the male blues and fins were making the regular low sounds. Since no one had ever found breeding grounds for fin or blue whales (whalers had sought such a gold mine for centuries), scientists began to suppose that such a place was not needed. Perhaps as the male whales called out for mates across an area of thousands of miles, the females who heard would then head toward the source of the sound.

Do they actually do it? So far there is only anecdotal evidence that whales themselves listen to songs from thousands of miles away. Serge

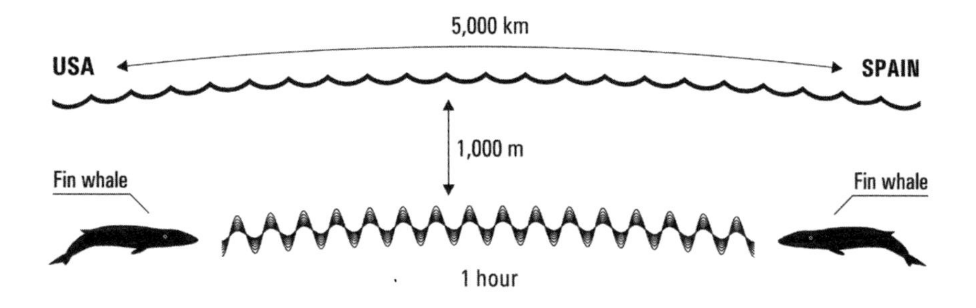


*Fig. 40. A low-frequency pulse produced by a fin whale could be heard by another fin whale on the other side of the Atlantic within the hour.*

Masse, a Montreal-based developer of cetacean research software (his latest creation is DC Dolphin Communicator, an app for Android phones, designed for two-way dolphin/human communication), remembers a Navy sonar man he knew who tracked fin whale booms in the seventies. Loud ones were detected right near his submarine, off Stellwagen Bank near Cape Cod, but there were very faint echoes that couldn't be placed. On the phone with colleagues near Spain, he got confirmation that whales off the European coast were making similar subsonic booms just an hour before. Why wasn't this published? "The information remained classified for decades," Masse smiles. "But now it can be told."

Scientists, though, have certainly heard baleen whale sounds from great distances away. With access to the Navy's super-accurate equipment, Chris Clark was able to track a blue whale for forty-three days from a thousand miles away. This giant blue whale sang continuously day and night. He began five hundred miles northeast of Bermuda swimming on a steady south-southwest course for three days. He passed just south of an undersea mountain and then turned toward the west and swam until he was two hundred miles northeast of Cuba. Then he turned right and ended up about a hundred miles from where he began. There he fell silent. Altogether this whale traveled twenty-two hundred miles over the course of a month and thirteen days.

Clark believes the whale may have echo-ranged off the seamount and then off Bermuda as a means of navigation. Even such deep sounds could be used for echolocation, especially if they are sung with such rhythmic precision, by an animal hip to long, drawn-out scales of time. Blue whale sounds at their source are 180 dB, as loud in water as a jet engine is in air. You wouldn't want to be listening too close to one. We would likely feel a huge rumble throughout our bodies if we swam nearby.

What is this long simple song then? A mating ritual, or a form of slow-mo sonar? Clark has shown that the mathematics for deep booming sonar could work, but there is no data to support that this is what is actually going on. But the thousand mile song takes hold of our imagination right away: "I could show you the evidence today. I can listen in Puerto Rico to a whale way up on the Grand Banks. Can the whales do that? You might well ask, 'What would they have to say?' Then you're suddenly putting on this silly human restriction. A whale might turn around and say to me, 'What would *you* possibly have to say to one another sitting just two meters apart?'"

Like so many great scientists, Clark is not afraid to be a bit of a dreamer. He is more concerned with saving the whales from increasing threats of noise and pollution than he is in figuring out what they're up to. More than once he has sought out the advice of musicologists: "Marty Hatch, a specialist in Indonesian gamelan here at Cornell, had this to say to me, 'You know Chris, you look at all this singing as *data*, but I think of it as a musical, emotional experience.' Musicians hear song, and this is where I sometimes lean away from the scientific and tend to agree with them. Why can't we just appreciate it as a phenomenon that is phenomenal?"

No human musician could stay in time counting as slowly as these whales do. These incredibly low thumps and moans are rhythms at so lax a pace that they are barely perceivable by human beings. Speed a blue whale song up ten times, and thirty minutes becomes three. Move the pitch up to the realm of a cello, bowhead, or a human moan and exactly every three seconds comes the same soft moan. Only at this slow sense of time do we hear the thousand mile song, a great sigh in the deep sound channel, echoing from one end of an ocean to another.

Is it right to call such a single repeating phrase a song? We call animal sounds "songs" for several different reasons. One, if we imagine a male is serenading a female, this begins to sound like music to human scientists. Two, if the patterns have musical qualities: rhythm, pitch, shape, and form. Three—and this to me is the most interesting and often overlooked reason—if the form of the sound does not affect the message conveyed, if there is no meaning to be explained apart from the singing, then we have a kind of communication that works like art, not language.

In 2007 several papers were published that shed new light on the deep dark sound world of the blue whale. Erin Oleson, with Sean Wiggins and John Hildebrand, discovered that blue whales off Southern California actually make three kinds of sounds, A, B, and D. The A and B sounds are quite consistent, sometimes heard separately, but more often combined into AB AB AB sequences, and these comprise what is usually called the "song" of the blue whale. The D sounds, in contrast, are very deep and variable downsweeps, as shown in figure 41. A and B sounds are generally made by males while they are on the move. D sounds are made by males and females, often when they are feeding. In bird terminology, we might label them "calls."

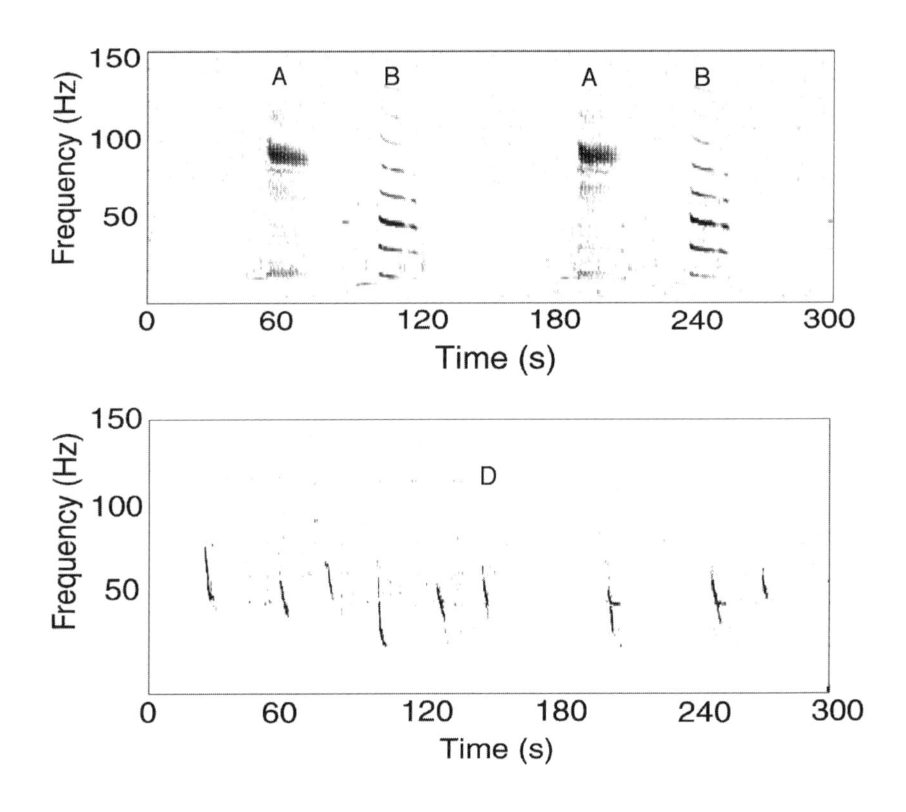


*Fig. 41. The A, B, and D sounds of blue whales.*

Using suction-cup tags attached to the giant whales, for the first time scientists were able to measure how deep the blues were while singing their ultra-low tones. Surprisingly, most of the songs were made at a fairly shallow depth. Figure 42 shows the behavior of one tagged male blue whale, diving deep and coming to the surface repeatedly. Note the shallow depth at which he made separated A and B sounds, and how he makes the transition to the AB AB AB song pattern, over a period of five hours.

He begins by milling about with another blue whale, whose sex is not mentioned. As he goes off on his own, our boy starts to travel northwest at four knots, making isolated A and B calls along the way. After two hours, he combines them and a song is born. Who is listening? We don't know. How far can it travel? Maybe twenty to fifty miles, nothing like the

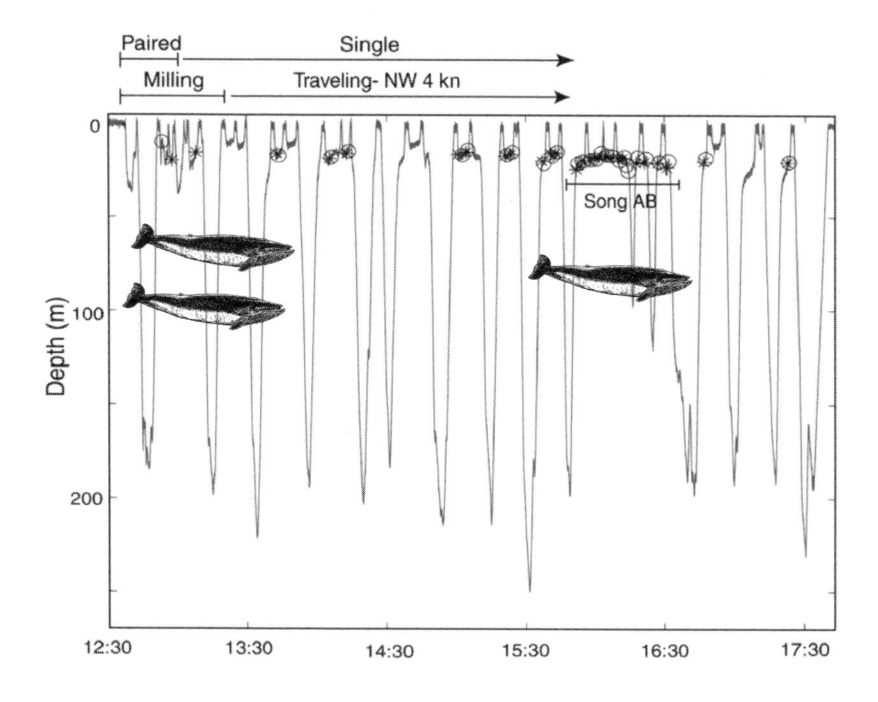


*Fig. 42. The rising and falling, singing and stopping behavior of one blue whale.*

thousand miles Clark and Payne believe to be possible. Further studies of tagged whales will be needed to show if the dream of distance is science or myth.

Another 2007 publication, by Mark McDonald, Sarah Resnick, and the director of the lab, John Hildebrand, is the first to identify clear regional differences in the song of the blue whale. Their paper argues that the songs are distinct enough to identify clear, music-defined populations of blue whales that are distinguishable by geographically defined dialects. Certain sounds in the song are sometimes blurred together, sometimes separated by a pause, but the same syntax always appears, unique to each part of an ocean. The authors point out that the song is much easier to depict in sonograms than it is to perceive by human ears, even if sped up and transposed to a higher octave. Even though each of the low moans takes several minutes to sing, the structure is far less complex than humpback song, so the distinct dialects are easier to see in figure 43:

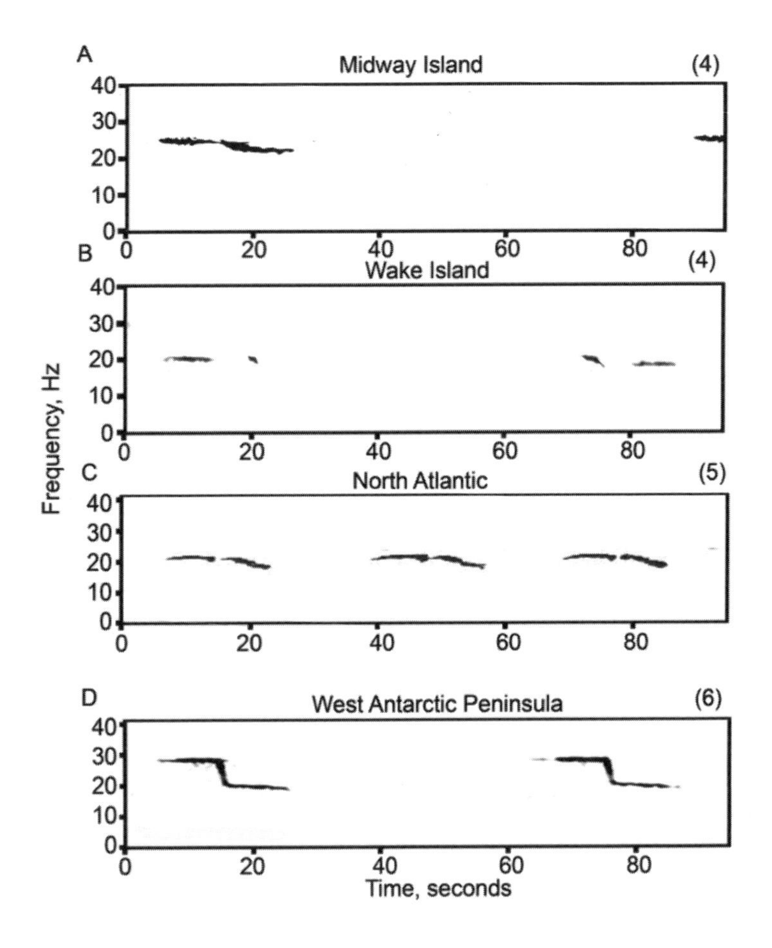


*Fig. 43. Blue whale dialects around the world.*

The differences are subtle, but important, because they enable researchers to identify distinct populations of blue whales around the world. Traditionally, separate populations of blue whales were distinguished within arbitrary International Whaling Commission boundaries, for the purpose of monitoring their number and the autonomous health of each group. But this paper suggests that the different populations now be reclassified along acoustic lines, kind of like giant acoustic clans of the kind Whitehead and Rendell identified for sperm whales. Except blue whales seem to congregate in geographically distinct acoustic clans, their movements differing from group to group, as shown in figure 44.

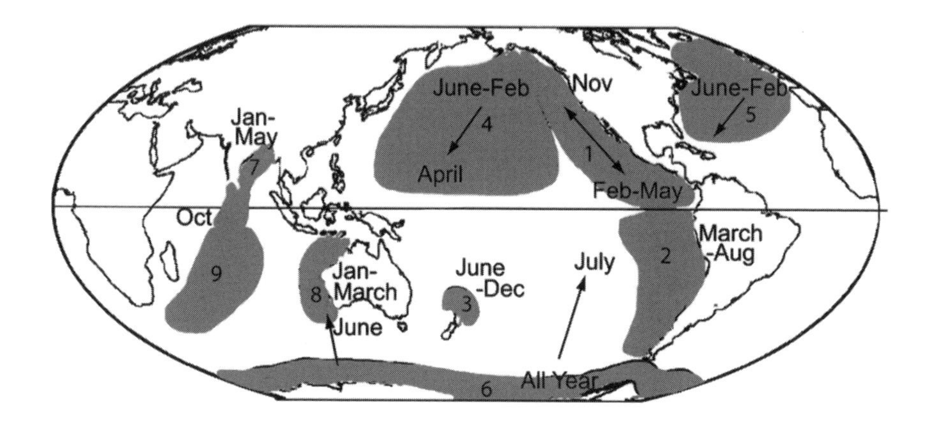


*Fig. 44. Where the blue whales are.*

Blue whales keep a rhythm too slow for any human to detect. And they may be getting even deeper. John Hildebrand has noticed that blue whale sounds are getting lower, going down an average of a tenth of a hertz each year. He believes this is because their numbers are increasing, and they need to communicate ever farther, and the lower the sound, the better it will travel. So far this remains an unpublished hypothesis. The largest animal that has ever lived on Earth makes a deep music whose subtle variations we are just starting to discover.

The ocean is by no means an opaque, silent world. It has always been naturally full of noise, at many frequencies. Wind and waves keep the sea rumbling at frequencies 500 Hz and above. Seismic explosions and earthquakes create sudden booms at the lowest end of the scale. In the tens of thousands of hertz and up to the hundreds, where dolphins squeak and hear, they have to reckon with the sound of the molecular agitation of water itself. Rain sends its high white-noise rush down beneath the surface. The underwater turbulence of the whole sea makes its presence felt at many frequencies.

But apart from the occasional underwater eruption, the bottom of the scale is remarkably free of sonic interruption. This fact, combined with the physics that allows very low pulses to travel so very far, may explain why these largest whales have evolved an ability to sing and to listen deep down in these frequencies. The deep sound channel, which allows 20 Hz sounds to lose just 1 dB in energy over two thousand miles, makes these tones the perfect pitches to send across great undersea distances.

Yet it is a much different sound world once human activity is brought into the mix. Today there are thousands of large ships coursing through global waters, most of them with big propellers making deep motor noises that peak around 90 dB in the frequencies between 20 and 70 Hz. This human-made sound is so ubiquitous in today's oceans at these very bottom frequencies that the long-range singing potential of these greatest whales is seriously compromised today.

Before propeller ships were trafficking all over the world's oceans, fin-whale pulses could easily travel a hundred and fifty miles at the surface and four thousand miles in the deep sound channel. With current levels of shipping noise, they are audible just fifty miles away at the surface, or six hundred miles deep down, before their volume is drowned out by background propeller rumble.

How much of a difference does this make to the whales? Birds in cities can handle the din; we've recently learned that they adapt by making louder sounds. In the acoustic space of the deep sound channel, can whales turn up to eleven? They could only hear a potential mate from a thousand miles away if the ocean were supremely quiet, and across most of the planet, that is not the case. Should we do something about this? Could we make our ships quieter for the sake of the great whales? Our noisy human world has changed the lives of whales forever.

Sudden human sound events in the ocean are much easier to stop. In March of 2000 whale researcher Ken Balcomb was astonished to find a Cuvier's beaked whale stuck in the shallows near his home in the Bahamas. Beaked whales are among the least known of cetaceans, since they usually live too deep in the water for us to see them often enough to learn much about them. Here he was, one of the world's experts on the beaked whale, face to face with one who was going to die unless he did something about it. It took more than an hour for Balcomb and his colleagues to coax this twenty-foot-long three-ton whale away from the beach and back into safer waters. The whole time the beast seemed confused and disoriented, but eventually he went on his way.

That was only the beginning. Soon afterward, Balcomb received a call that five beaked whales had been stranded on nearby beaches and most were still alive. By the end of the day, fifteen beaked whales were writhing on Bahamas beaches, along with two even bigger Minke whales. By the end of the day six of the beaked whales died. What happened? The U.S. Navy had been conducting a mid-frequency sonar exercise, bombarding the channel with reverberations around three thousand Hz, to a level of 150 dB, every twelve seconds, for sixteen hours.

Subsequent autopsies found that the whales had died not so much because of hearing loss, but because they had developed a kind of decompression sickness (a version of the bends), because the particular frequency used by the Navy matches the resonance frequency in the air space inside their heads. Balcomb describes it as follows: "Envision a football squeezed to the size of a ping-pong ball by air pressure alone. Now envision this ball compressing and decompressing hundreds of times per second, between your two ears. This is what the Cuvier's beaked whales experienced as a result of the Navy's sonar testing in March 2000. Airspace resonance phenomena resulted in hemorrhaging, which caused the stranding and deaths in the Bahamas."

Although at first they were reluctant to admit complicity in the deaths of these little-known whale species, eventually the Navy did accept blame for the whale strandings. In theory they have agreed after years of negotiation to refrain from testing mid-frequency sonar in areas where whales are known to be present; however, in practice these sounds could travel up to a hundred miles and still have enough energy to harm whales. Sonar testing has been linked to whale deaths in about thirty instances over the last forty years; more than a hundred whales have died. The Navy defends itself by saying that this number pales beside the amount caught in fishing nets or cumulatively harmed by pollution, but sonar testing is much easier to do something about than these more widespread threats.

In early 2006 the U.S. Marine Mammals Commission released a report on acoustic impacts that included contributions from all the major stakeholders in this debate: an environmental caucus, a scientific caucus, an energy industry caucus, a shipping caucus, and a federal caucus. It is a fine case study of how an issue can look completely different depending where you're coming from. The environmentalists said if there's even a chance that an underwater sound might harm a whale, we shouldn't make that sound. This is the famous precautionary principle—what might be bad could be bad; do nothing unless it's absolutely necessary. The energy industry emphasized its need to fire underwater air guns and unleash explosions for seismic exploration, because of course we need the energy, and of course they need the money. Scientists said we simply don't know enough to be sure about the impacts, so we must do more research. They suggested that we really should study the continuous impact of moderate noise more fully, but they did not suggest how this should be done. The shipping industry agreed in principle to investigate reducing propeller noise, especially if the change could be legislated gradually, and if convincing proof were provided that general shipping noise was a problem for whales.

Since this report presented so many different viewpoints on the same problem, the committee was unable to reach a consensus. In the summer of 2006 the Navy was preparing to conduct exercises in the Pacific that involved mid-frequency sonar, the kind known to harm beaked whales. They had already agreed to certain restrictions on the tests: no active sonar in deep canyons where beaked whales might be hanging out, and sonar pinging would be turned off whenever there was a possibility of whales in the vicinity. There was still protest from environmental groups, but the Navy insisted the tests were essential: "We need to track the submarines of our possible enemies, nations like Iran, North Korea, or China."

The Natural Resources Defense Council immediately sued the Navy with a cease-and-desist order, and a judge in the Los Angeles District Court agreed that going through with the tests would violate the National Environmental Policy Act. The Navy agreed to alter the testing site somewhat, pledging to steer clear of the recently decreed Northwestern Hawaiian Islands Marine National Monument, instructing all personnel to be especially alert for whale sounds. No longer were these voices mere "biologicals." They would now be an important part of a sonar operator's listening field. Environmental lawyer Richard Kendall called the settlement "a significant step forward in the protection of our oceans," while an admiral called it "a small number of additional mitigation measures."

It looked like a rosy compromise until January 2007, when the Navy suddenly claimed an "exemption" from the Marine Mammal Protection Act and gave itself permission to conduct whatever tests it wants for the next two years, citing national security as their justification. So much for the image of a rare whale helpless on the beach, pulverized by resonating tones inside its air sacs, all the result of secret tests of a technology designed to flummox enemies we're not even sure exist.

The Navy still finances much of the research on whales and dolphins. They did turn their classified underwater expertise over to those scientists who could best appreciate it. They remain an ambiguous influence on our quest to understand the sonic abilities of cetaceans. They do not feel the issue is resolved and are still seeking advice on how to remedy the situation. Rumor has it they've even asked the endlessly creative Jim Nollman for assistance on this problem.

The rise of chronic human noise in the oceans remains a more worrying threat to whales, because it is much harder to control than the occasional sonic blast. As Peter Tyack describes it, "The deepest whale songs used to carry for hundreds of kilometers. Now it turns out that the thing that dominates the frequency range down there is shipping noise. This is where we put the most acoustic energy in the ocean. If that increased low-frequency din is preventing a male and female from finding each other, it is important. It's not as visible as a dead whale on the beach, but it might be much more prevalent."

A recent study shows that North Atlantic right whales, of which only three hundred remain, have started making higher-pitched calls over the last half century. This change may be a response to a habitat full of more big propeller ships than ever before. Each decade the hum of motors adds nearly 10 dB of rumbling sound to the background noise the whales have to reckon with.

We can't be sure of the precise effects of chronic loud human noise on the whales, and this uncertainty is capitalized upon by those whose activities would be hampered by having yet one more problem to worry about: seismic explorers looking for oil, the Navy testing new equipment. Their position remains, "If we're not sure it will harm the animals, why worry?" Environmentalists and whale lovers favor the precautionary principle, where what we don't know can still hurt us, quite a bit. Better to be careful, and show greater respect for these great singers of the seas.

I tend to agree with Jim Cummings, founder of the Acoustic Ecology Institute, a now defunct clearinghouse of information on all sonic assaults on our environment. After carefully reviewing this issue for many years, this is what he has to say: "There are other foundations to stand upon in creating public policy than those built of scientific certainty. We need to remember that the effects of our actions tend to ripple and interact in ways that we cannot predict, or often even fully recognize. This is surely the case with making extreme noise in the sea."

Science must praise uncertainty, as it progresses by constantly scrutinizing itself and questioning assumptions, scrupulously amassing ever more data. When it comes to the world of whales and the remote deep sea, its conclusions are understandably incomplete. Based on its own criteria for objectivity, science can rarely tell us exactly what to do. So we should not be misled into thinking that our care and feeling for whales and their world is something subjective, emotional, and easy to discount when it comes to planning their future. Our species has no future on this planet until we consider the flourishing of all other forms of life as part and parcel of human progress. There is a gentler, more humane way to live along with the sea. We must learn to hear the ocean's music, so as not to play our human part too loudly.

# 9

## NEVER SATISFIED

### Getting Through to a Humpback Whale

"What do you expect to accomplish with this?" Mark Johnson glares at me. He's a grizzled New Zealand engineer with a ponytail who works at the Woods Hole Oceanographic Institution. "Why would it surprise you that you got a response? These are acoustically active creatures. I don't see the point of fucking with animals just for the hell of it. I don't believe in diving or swimming with whales either."

We're speaking at the American Cetacean Society annual meeting at Ventura Beach. I fumble with an answer. "I . . . just want to see what happens."

He shakes his head and squints at me. "If you're trying to learn something about whales which could then translate into a tool for conservation, that's worth doing, but simply trying to enjoy an event in which the animal doesn't even know he's participating, to be honest, I don't see the point. It doesn't really yield anything except a gratuitous level of self-satisfaction."

The interview was not going well. I try to defend myself. "Mark, I'm trying not to pat myself on the back. Music is knowledge too. I want to learn something, and make some interesting sounds, sounds that can't be made by one species alone."

"What's the point if you haven't got a hypothesis?" he shouts back.

"I do have a hypothesis," I tell him. "But it's musical, not scientific. My idea is that there's more to music than humans can make on their own. A wandering jazz musician could enter a club in Tokyo or Istanbul, climb on stage, and right away join the band. You listen for a familiar sound, some rhythm or chord changes, and try to join in, finding a way to a common music. This is how you make music together with those with whom you cannot speak."

Through music you can cross cultural lines, not because music is a universal language, but because it is a fluid and emotional form of communication, able to embrace and make use of unknown sounds. Something new can be produced between people from different sides of the world, or even between different species. Music may even be useful to science in its quest to understand the vocalizations of whales.

Jamming with humpback whales could be the ultimate interspecies experiment, because they are the animals with the longest performances we know. The complexity of their music is greater than that of any other whale, and it refuses to stay the same. We know they are interested in new sounds because they change their songs every year. We know they are able to learn from experience because, as Jim Darling has shown us, when a non-singing male joins a singing male he picks up the gist of the new song and then goes off silently to internalize it, as if to prove he is with it, hip to the season's latest hit.

Sure, they respond to each other in this special, ritual way, but will they respond to me on my squeaking clarinet? Scientists are most interested in how animals respond to controlled stimuli, because it helps them ascertain what differences in sound they can discriminate, and which aspects of song matter to them as opposed to those that matter to us.

These tests are called "playback experiments," because recordings, usually of the animals' own sounds, are played back over tape or other recorders. By doing such experiments with male songbirds, we've learned all kinds of things about how they respond to other males, other species, and artificially adjusted versions of themselves. However, such experiments have to be designed very carefully, because animals are smart. It doesn't take long for them to figure out that a recording is not a real animal. They habituate to the sounds, and they get used to the experiment. It becomes more of a game than a test of how they naturally behave. More often than not, they soon lose interest.

In the past, attempts to play humpbacks back their own songs have not provoked interesting responses from the whales. Scientists have had better luck with the interactive social sounds more characteristic of summer feeding than winter breeding. When presented with these more grumbly noises, the male whales sometimes charge at high speed toward the boat with the speaker, just diving under it at the last possible minute!

Peter Tyack, who conducted such an experiment in 1982, says that in all his years observing humpbacks, he has never seen a whale charge a boat except when he was playing back a tape of social grunts, clicks, and burps—sounds usually heard only in summer. When he played the winter breeding song back, however, fourteen out of the sixteen groups of whales who heard it moved away.

Three years later Louis Herman and colleagues tried a more elaborate experiment, with 143 playback sessions. They noticed a lot of approaches by single whales or adult pairs, but again, not when the song was played. Twenty-two percent of pods approached when they heard a feeding sound that had been recorded the previous summer in Alaska. Eight percent approached winter Hawaiian feeding sounds. Three percent approached the playback of a local winter song, and 4 percent approached the playback of ten seconds of "random continuous variations of pure-tone frequency," a sweeping woozy note played by a Wavetek synthesizer. No whale approached a silent boat.

Although these results were not encouraging in terms of getting whales to listen to our imitations of their sounds, these data did turn out to be useful in practice. In 1986 one wayward humpback whale swam up the Sacramento River in California and nearly got stuck. He was nicknamed Humphrey, and he kept heading upstream, where he would most certainly run aground if he kept going. Fifteen marine mammal scientists got on a conference call and tried to figure out what to do.

With Tyack's results in mind, they said, "Let's get a boat out there and try to play him some humpback social sounds. That has the best chance of getting an approach." The famous nature recordist Bernie Krause got involved and offered some sounds from his vast library. Sure enough, when he heard the feeding call off a ship, Humphrey immediately turned around and followed the boat fifty miles in seven hours downstream to the Golden Gate Bridge, usually staying no more than fifty meters from the source of the sound. They kept going the next day and Humphrey found his way out to sea. During the next summer Humphrey was seen frolicking around the Farallon Islands in the Pacific 40 km west of San Francisco. He was positively identified by markings on his tail flukes.

In the spring of 2007, two more humpbacks, this time a mother and calf, got stuck up the Sacramento River and stayed in the California news headlines for weeks, as scientists and marine managers struggled to find a way to encourage them to head in the right direction before they would be damaged by too much time spent in fresh water. This time recordings of feeding sounds had little effect, but the whales, nicknamed Delta and Dawn, eventually returned to sea on their own. Playback is no sure way to move an animal about as big as a city bus.

Johnson still looks skeptical. I quote from the history of science to him. "Other species have other senses of beauty and form. Even Charles Darwin knew that, although biologists seem sometimes to have forgotten. He wrote in *The Descent of Man* that the beauty in nature was effectively chosen by the animals themselves, over generations of female preference. They liked certain traits, and those were passed on. Each species has its own aesthetic, we have to stretch our ears to get a grasp of whale music. Besides, showing people how beautiful whale songs are is what really got us interested in saving them."

"Right," he still glares back, "but no one had to play guitar to a humpback to do that. I have every respect for passive techniques. The only reason to fuck with something is if you have a well-worked-out hypothesis, and a good measurement technique, so you would be able to reasonably conclude that the animal is responding," he sighs. "But a musician might say his expression would be inhibited by such a controlled situation, so perhaps these approaches are incompatible. But for God's sake leave them alone unless you have a really good reason to bother them! Imagine poking an animal with a stick!"

"Jamming with a whale is not the same as a spear in the ass."

"How do you know that?" Mark fires back.

"Scientists, not musicians, are the ones who shoot darts into the backs of great baleens," I remind him.

"Sure, most scientists do a terrible job when designing experiments. They're mutants!" This engineer is a bit agitated.

Johnson himself is known for inventing the D-Tag, a small digital recording device that can be attached with suction cups to the backs of whales and dolphins. On board are tiny stereo microphones and sensors that pick up orientation and acceleration. All the information is recorded onto tiny flash memory cards like those in digital cameras.

After a few hours, or at best, a few days, the recorder falls off the whale and floats to the surface, where it is located by radio transmitter and brought back to the lab for analysis. No whale has been cut, bruised, or harmed in the study.

Mark is justly proud of his invention, which he's designed to aid the scientists and help them learn from the whales with less harassment. The data generated from the D-Tag have enabled us to learn how whales are moving when they make sound, and it's been able to record the previously unheard sounds of the elusive beaked whales, the species most susceptible to mass strandings induced by mid-frequency military sonar testing. When talking about these animals Johnson's anger subsides and his heart seems to shine through:

"Beaked whales, I'm really in love with those animals, I want to save them. Now, if it weren't for the strandings that we need to explain, I'd say, leave the damn things alone! These whales are doing just fine without us. The last thing we need is a bunch of scientists interfering with them for the good of science alone. However, when things are found out about their behavior that help to change people's perception of these animals, public opinion on them improves. My bottom line is, the research must fit into the goal of conservation."

This is really a rather recent goal for field biology, which until several decades ago tried to keep its distance from more political aspects of ecology. But now we know that a vast number of the Earth's animal species are in danger of extinction, so we all must fight to save these creatures before it's too late. Johnson's work has helped to prove that Navy tests do real damage to these elusive whales.

Before the D-Tag, the sound world of the beaked whale was completely unknown. "We didn't know if they made sounds at all. With the tag we found they make exactly the same kinds of echolocation clicks while foraging as other toothed whales. But then they make a little sweep, a mini sonar, like a very compressed bat sound, nothing like sperm or pilot whale sounds. We've recorded this whole new style of click in at least two species of beaked whale.

"Humans can't hear this kind of click at all, because it's so high pitched, but you can hear the sound that a *tagged* whale makes, because you hear the sound conducted through the bones of the whale right under the skin! If you play the sound back at a quarter speed, slowed down, then you'll hear all the *other* beaked whales around. We get better recordings of the other whales than the whale who is carrying the tag! As Ken Norris said, 'Every sound a whale makes is a social sound.' You could argue that every sound they make is an echolocation sound as well."

Now that I had got him genuinely excited I asked Mark why he thought humpbacks changed their songs from year to year. "I'm a humble engineer, I don't know. If you're asking what an animal does, that's something engineers can figure out. If you're asking why, that is beyond our expertise. But I would guess they just like variety, same reason we shuffle tunes on our playlists. We're not all that genetically different from these animals you know, even though evolution has got us there on radically different paths."

"So you don't feel qualified to comment on whether or not whales are intelligent?"

"What's misguided is when we call animals *un*intelligent. The trouble is that our idea of intelligence is so damn limited that we can only see it when the animal has some great ability, or if it interacts. One could argue that it is a misguided or dopey animal who interacts with us—not a smart move to play music with *you* or get too close to a whale-watching boat. You could end up being chopped up by the motor like that poor orca Luna."

"Come on, Mark, making music with whales is not the same as running over them with a powerboat."

He's definitely angry now. "Who gave *you* the right to mess with these animals? If you can't explain it, you shouldn't be doing it. Don't go telling everyone to go out and play music with the whales to share their own interspecies communication fantasies."

(*Fucking with animals, fucking with animals, fucking with animals for the hell of it...* This refrain swirls in my mind for several years until I use fragments of this interview in a piece on my album *Whale Music Remixed* in 2009. Guess such a track would still be bleeped out on commercial radio.)

I still believe in my plan. "Listen, none of my underwater music is too loud. I'll play a bit, then stop. Listen for a reaction. Try it again. I'll respect the whole experiment as a musical situation, with respect for the whales who might ignore me, might join in, or might ram the boat. Then I'll definitely stop. It's much less of an imposition than those loud warning sounds you played to right whales off of Cape Cod. Wasn't that a lot of noise for those poor animals, a species of which only three hundred remain?"

"Well, that experiment is done, it's been repeated, and that thing bugs the shit out of me. Out of a controlled exposure experiment, you only get funding to do another controlled exposure experiment." Johnson shrugs. "At least one important thing came out of that project. We learned that those right whales come to the surface after hearing a loud alarm sound, exactly the worst place for them to be, because they might get hit. There was a lot of talk at that time about putting alarm sounds on all ships, but now that idea's dead because we know it traumatizes the whales too much. It would have been a very bad solution, terrible for the whales. What might have happened is they would have habituated to the sound very quickly, and the alarm would become like crying wolf."

We head back to the main hall. The next talk is by *National Geographic* writer Doug Chadwick, who is regaling the crowd with some of his adventures crisscrossing the globe with some of the greatest nature photographers and the largest budgets in the magazine world, when he starts to challenge the scientists right here on their own turf.

"Whale science is truly hard work, and there's no one I respect more than these intrepid researchers who go out every day hoping to observe the whale behaviors that they're focusing on, meticulously collecting data for projects that take many seasons to assemble, and many more years to analyze. Their dedication to the unpopular cause of decoding complex animal behavior deserves our deepest admiration." Now Chadwick pulls a fast one. "So far, scientists have been rather timid in their approaches. They observe without trying to disturb. That is the classic way of science but with animals this magnificent and seemingly intelligent, it's time for something more. We must *interact* with the animals, engage them, take them more seriously as individual thinking, caring, and moving beings."

I smile. A thumbs-up from a fellow traveler, telling me to get right out there and jam. Perhaps my dreams are not so lonely.

"Right," Johnson grumbles. "I bet you like *that* point of view. Now that we've got all that out of the way, let's go get a pint."

Luckily I found other engineers more sympathetic to my plans. I first discovered the Whalesong Project while surfing the internet looking for "whale" and "song." They're an outfit that broadcasts live whale songs from Hawaiian waters twenty-four hours a day during the January–April mating season, when the seas are full of whooping, howling beasts. It's the brainchild of Dan Sythe, an engineer who owns a company that makes radiation detectors. He used to run the business out of Sonoma County, but when he first heard humpbacks about ten years ago, he knew he had to change his life.

His upcountry office is full of all kinds of audio recorders and every manner of electronic gadget. The ocean is visible past the eucalyptus forest below. "The old Hawaiian songs have *kaona*, secret allusions, metaphor, sometimes playful, sometimes with deep, mysterious and meaningful expression of emotion," says Dan, a tall, tanned guy with boundless enthusiasm. "The whales seem to have these too: *Polinahe* some parts soft and gentle, *nahenahe*, sweet and melodious, and of course *'i'i*, that deep, raspy, growly quality. Whales appear to have an easier time at that than most of us *po'e* do. They seem to express their emotions with complete freedom."

In the late nineties, Sythe moved to Maui to figure out how to let the whole world hear these beautiful songs at the moment they are sung. "Our business is flourishing, and it's still located in California. I don't deal with the military, only projects that protect people, creatures and the environment," he assures me. "I don't think the way some people in government and business do. They want this country to be afraid; they're in the business of scaring people. One of our biggest customers is the New York City Fire Department. Our devices can detect alpha, beta, and gamma radiation, much more sensitivity than the federal government requires for first responders." He proudly hands me his latest design. It is rubbery, rugged, and functional, made in the USA. So what does this have to do with whales? "Protecting the environment should be a primary mission of our country's efforts toward defense and homeland security. Without healthy oceans, how can there be any security?"

Seems like Hawaii is a fine place to labor at the fringes of high technology. Dan's partner in the Whalesong Project, Kent Noonan, is another engineer living high on the slopes of Haleakala. As a day job he's an inventor, currently trying to build the fastest switch ever made using fiber optic technology. In his spare time he farms coffee and bananas, creates mystical electric field machines that some claim can cure cancer, and experiments with stereo hydrophones dangling off each side of his ocean-borne kayak. "I just like to figure out what can and can't be done," says Kent, resembling a modern-day magician with shoulder-length white hair and a trimmed beard. "If it is possible, I will build it."

*Fig. 45: David Rothenberg and Radha Ala visiting the Whalesong Buoy in 2010*

But both of these fine minds are wrestling with another problem: how to broadcast the sounds of humpback whales to the world over the internet twenty-four hours a day throughout the breeding season when the whales are a-singing. They've set up a big yellow buoy named *Hokumoanalani* [Star of the Heavenly Ocean] in the bay off Kihei that broadcasts remotely via a radio signal to a Macintosh in a shack just off the beach. This machine streams these whale songs across the internet for the world to hear at www.whalesong.net.

"These sounds are important, and beautiful," says Sythe. "The whole world ought to be able to hear them, live, right now, as they are being sung. I've spent my whole life savings making this possible. It just seems like the right thing to do." Click on the site yourself, anytime between January and April, and you'll hear humpback whale songs live as they are sung, direct from the crystal-clear waters of Maui. Maybe there are distant wails, maybe a nearby screal, maybe a solo male singer giving his all with the whole fabulous tune of the year. Maybe nothing save a few power boats. Maybe a mischievous human swimmer moaning into the hydrophone. What's remarkable is that this technology works! You can tap into this humpback music from wherever you are.

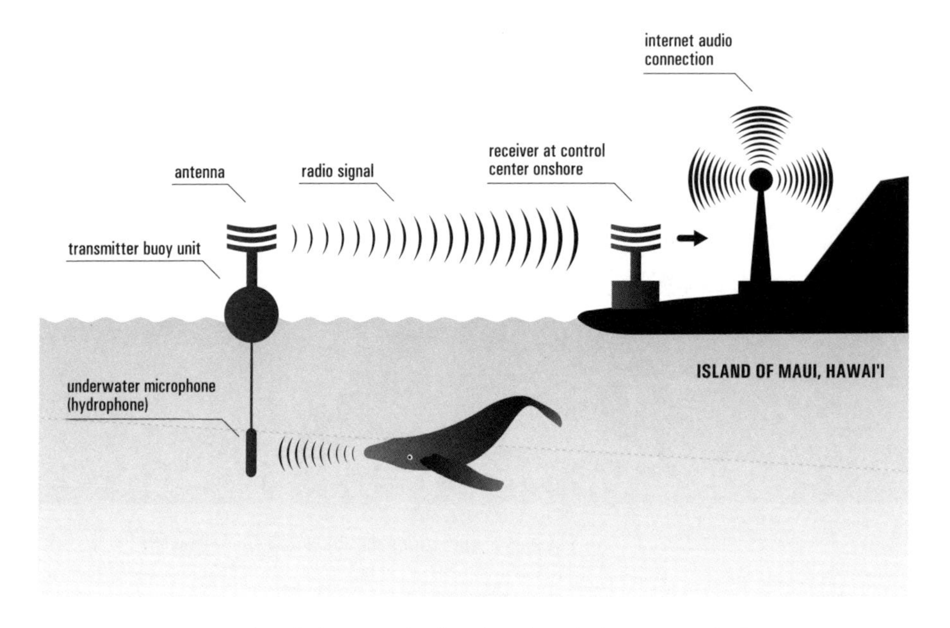


*Fig. 46. How the Whalesong Project broadcast humpback songs over the internet.*

Dan and Kent are my main first contacts on Maui, and they are quick to point out to me that it's not going to be easy to find someone who will let me jam along with Hawaii's whales. You can't just go to the waterfront at Lahaina and sign on with one of the many fine whale-watching tours announcing that you want to play some music with the whales. It's still against the law to torment or annoy a marine mammal in any way that might upset behavior patterns. And this of course includes playing any sounds that might cause the whale to alter what he's doing: exactly what a musician like myself would like to hear happen! I want him to change his song, if only for a moment or two. It's also illegal to swim in the wild with whales, something many adventurous tourists dream of doing. No one is allowed to deliberately approach within three hundred yards of a humpback whale. That's pretty far away.

Scientists operating under federal permits have license to do all these things, in the name of research, when trying to find out what is going on in the world of whales. Only people listed on the permits can go out with scientists, and I'm fortunate that Jim Darling has agreed to put me on his permit this season so I'm able to go out with his team. Since his latest research involves playing back sounds to the whales, mostly their own songs of course, he's agreed to consider my clarinet playing as a "control sound," a term in playback research that means some presumably random or unimportant intervention that is different from the main material that is going to be played back to the animals.

But meanwhile, there are any number of rogue whale safaris going on, unadvertised, through-the-grapevine trips that play by their own rules. They don't want to harm whales either, but these people have their own approaches. Many have spent years on the water with whales, and have a lot of experience reading their moves, not through analyzing data, but through time and attention on the water. Some claim a telepathic connection to the animals and have their own methods to prove it. Others just want to sing and celebrate the beauty and passion of these giant sirens of the ocean depths. "Come Join the Spirit Sailing Journey to the Cetacean Nation," says the e-mail invite, full of tiny print listing all the fabulous love and music that will be on board. "Leave your shoes on the beach, swim out to our catamaran, drums, and shakers will be on board. Our guest musicians will play cellos, harps, and guitars. We'll have flippers, you can dive in with our giant friends." As soon as we leave the harbor, clothes tend to come off. A woman named Mahana, who was once called Lauren, waves and shouts the old Hawaiian word for humpback out to the sea. "*Kohola, koholaaa*, we love you, come close to us, we're waiting." The passengers start to sing and dance, the warm sun beats down on us, and we see the first great humpbacks cavorting in the bay. It's prime mating and calving season, and they're everywhere. A few hundred yards offshore and we're all mostly naked, trying to tune in to that great whale energy.

The old Hawaiians called the humpback *kohola*, and they even named a nearby island Kahoolawe, because it looks so much like the back of a whale. But there are no ancient myths about whales, and the islanders never hunted them or talked much about them. Revisionist island history has come up with lots of tales about whales, but where were they long ago? Some say there just weren't as many humpbacks cavorting in these waters until recently, because whalers were plying these waters as early as the 1700s. Others say there were plenty of Hawaiian myths about the animals, but they are too secret to let us *haole* find out about them. Why let us invaders have the best stories?

I believe it is a question of attention—nobody noticed. They may have been diving, but they didn't think to listen while under water. Yet when our catamaran drifts so close above a singing male that the hull is shaking with the low *'i'i* tones easily audible above water with no amplification at all, I have to wonder. Surely someone detected this in centuries past, when we humans were either pursuing the beasts in tiny boats armed with long harpoons, or telling stories about how our ancestors traveled across the deep blue seas?

Maori writer Witi Ihimaera wrote the famous novel *Whale Rider* while he was living in New York in the 1980s. A humpback whale had been sighted in the Hudson River, and he was inspired to resurrect a little-known Maori tale about one ancestor arriving to Aoteoroa from Hawaii on the back of a whale. The novel is full of rich descriptions of singing whales: "Suddenly the sea was filled with awesome singing, a song with eternity in it, a song to the land." In the story the whales sing because they mourn the loss of their human rider, who used to play a flute to them. "You have called and I have come, bearing the gift of the Gods."

Powerful stuff, but the original Maori myths say nothing about whale songs, sometimes even emphasizing the sheer *silence* of the noble leviathans instead. There would never have been such *Whale Rider* language without the famous recordings of Watlington, Payne, and McVay. Now that we humans know about whale songs, we imagine we have known about them all along. But the painful truth is that, although they can be quite easy to hear even without all this technology, no one bothered to listen to them until they knew there was something to listen for. We remain prisoners of our expectations.

One Maori myth about the bond between humans and whales centers on revenge: the great chief Tinirau was so powerful he even had his own school of whales. He could call them from shore, and they would come near to the beach and play, cavorting around in circles and gushing spray through their blowholes. His largest and favorite whale was Tutunui, upon whose back Tinirau would climb and ride blissfully through the waves, like a surfer on the ultimate board. Tinirau always felt safe on the back of Tutunui, his greatest whale.

One day the priest Kae came to visit to perform a ceremony, and Tinirau really wanted to impress him. "Try this special treat," he told the priest. "Tutunui, come here," he called out to the ocean. The giant whale obeyed and came very close to the shore. What did Tinirau do but hack off a small chunk of his flesh! "Take a bite," he smiled to Kae, who looked concerned. Tinirau reassured him: "What are you worrying about, this whale's so big, he won't miss this tiny bit."

Kae had never tasted anything this delicious, and he wanted more. "Hey Tinirau," he grinned. "Mind if I borrow your whale for a few days? I need to cruise out to some outer islands and do a little business. That big guy looks like a good way to travel." Tinirau was unsure about such a loan, but it was more important to be generous to a priest than express your inner doubts.

"All right," he said, "but you must be very careful, especially close to shore. This whale knows when it gets too shallow, he'll give a shake and then it's time to jump off. If you stay on his back he'll go right on up on the beach and dry out 'til he dies."

"I get it," nodded Kae, thinking sinister thoughts. "I'll do nothing to harm him."

But Kae rode Tutunui straight back home to his village, and he saddled right up onto the beach where Kae's fellow villagers watched him dry out and die before hacking up his flesh and having the greatest meal of their lives.

The scent of barbecued whale carried so far that it reached all the way back to Tinarau's house. He knew the smell and he wept with grief.

It took a while, but Tinarau's men hunted Kae down, blindfolded him, and brought him back to the chief. When they took off the cloth and Kae saw where he was, Kae hung his head low and began to wail. When he again looked up, Tinarau stood before him brandishing a large club.

"Did Tutunui make so loud a noise when you slaughtered him?" said Tinarau just before he brought the club down on Kae's head, killing him instantly.

And so was Tutunui the pet whale justly avenged.

I'm sitting on the boat, feeling the whale rumble under me. He is far from silent, something the Maori never knew. But I too want something from the whale, and don't want to be quiet. I toss in the hydrophone to get a better listen. It's an incredibly loud solo male song.

We plug it into an on-deck speaker so all can hear. He's so close it distorts our system right away, sounding like some Jimi Hendrix of the sea. When we drift away a bit I get a cleaner signal. I take out my clarinet and try to play. No underwater speaker today.

"*Ooh*, beautiful man. Relax bro. He can hear you, the whales are telepathic, dude!" A dreadlocked guy holding a big fat book, *Songwriters on Songwriting*, looks up from his notebook where he's writing his own new song. "Check out these lyrics man: 'There's nothing so beautiful as a whale under the sea, except my new love, lying on top of me. All *right*!'" A naked woman starts to blow a didgeridoo toward the water right next to my ear. I remember Nollman in the Mediterranean, what a hard time he had concentrating. But it's hard to worry about anything in Hawaii. I'm playing high, detached notes, basking in the sun up on deck, leaving space for that whole whale song. Is he grooving on the didg and the reed? We don't know what they hear, what they sense.

Some people are listening to the music, some are ignoring it. Like your average jazz club, except the sun is shining, we're out in the waves, it's the middle of the day, everyone's all smiles and no clothes. A guy with a long gray beard comes up to me and says, "You know, I've been on these boats before, and usually everyone just chants and sings their own shit. But you're actually listening. The clarinet is one of few instruments that merges with these whale sounds. Your music does fit in with them."

I have no idea if he is right or not. The encouragement counts a lot, because I'm often doubting what I'm doing here. Why would the whales want to listen to me? It's a crazy hope, to make music with animals with whom we cannot speak. Easy to agree with my critics who say I'm just full of myself, tooting my own horn. Somehow I just have to keep doing this. It interests me. On the course of this journey it has become what I do. Maybe I'm finally learning how to do it.

The next day a big storm blows in from the west, the backward-blowing Kona Wind, and the water is off limits to whale listening for several days.

*Fig. 47. Break out the didg.*

Fifty-mile-an-hour winds at the Haleakala summit spray lava dust in my face. On the shore the sea is fine for expert parasurfers but the rest of us must stay on land. A sailboat lies overturned in the Lahaina harbor, and tourists let the moist winds fly onto their faces.

I decide to visit Flip Nicklin, the great whale photographer, famous for his work in *National Geographic*, who's also waiting out the storm. A great bear of a man who grew up in Hawaii perusing these waters, he's all smiles and experience. I ask him if he ever heard whales growing up. "That's a good question. I was diving here in the sixties. We didn't *think* about listening for whales. We weren't listening for anything." His father Chuck Nicklin was one of the very first underwater whale photographers, working on all the earliest *Geographic* whale stories including the 1979 feature that included the ten million copies of the sound page record of whale songs. Since then Flip has taken up the family trade, and his images are the most well known of all whale portraits. They grace the pages and covers of nearly every book on whales.

After taking photographs to aid scientists in their exhausting research, Nicklin has lately become a bit of a scientist himself. He's one of the coauthors of Jim Darling's landmark recent paper on what really happens while a male whale sings. But Nicklin doesn't like to talk much about that work, and he doesn't want to take sides on whale-related issues. He's traveled with whale hunters, whale savers, and whale cops, and his work depends on the fact that everyone accepts his presence. "I don't like to think for people," he assures me. "I want to put images out there to give people something to think about. That's enough for me."

Flip has had experiences few of us have shared, and he's not even allowed to reveal them. "You don't really know a whale until you've looked one in the eye. Under our permit we're not even *allowed* to show you how close we get to those whales. Every image we publish has to be reviewed by the Marine Mammal Protection people. That's not true for people who get close to whales who are not under scientific permits, but we have to follow those rules if we want to keep doing our research."

His condo is full of visiting photographers. Cameras are piled all over the place in watertight boxes. Assistants are downloading images and cataloging them on computers. Wetsuits are drying off on the porch. Everyone hopes the weather will soon improve.

"I don't work with anyone until I learn to trust them. People say all kinds of things about whales, you have to know where they come from, what they have to back up their claims. I've learned to be cautious and careful, skeptical of what people claim they know or they've seen. The more sure someone is of what they think the whales are doing, the less I trust them. The best guys, like Jim Darling, are deeply uncertain, because they've seen so much."

Nicklin has spent many years close to whales, and he presents a humility greater than anyone else I've met: the more you see and hear of whales, the less certain you should be about them. Do not trust he who claims a special connection to the *kohola*. Follow those who admit we know almost nothing about them at all.

On Super Bowl Sunday, the sun is shining. Still, there's hardly anyone out on the water. I can't believe they're all home watching TV! The ocean is as smooth as glass. "Mother and calf, at ten o'clock," announces Willy the Whale, pointing to the port side off his agile Bayliner. "She's fifteen years old, and the baby's four weeks."

"How do you know?" I wonder.

"Dowsing. I learned it a few years back. It's extremely effective with whales."

Willy's full name is William Bennett, and he once worked as an engineer on the Hubble Space Telescope in California. Now he runs a solar water heating lab at Maui Community College when he's not out dowsing for whales. He can sense the ages of the whales, their sex, and where they are going. Sal Cerchio knew him back in Santa Cruz, where he once looked into the eyes of a beached pygmy sperm whale and recorded its dying thoughts. Another time he dowsed the fleas off his dog.

We're joined on the boat by former whale biologist Heather Harding. She used to be a research scientist and now has a more intuitive bond with the subjects of her research. "I wanted to be a marine biologist since I was five years old. But I always considered myself a spiritual person." She is the one person I have met who has tried to think philosophically about what it means to be a field biologist, and she has ended up stretching the boundaries of her work in a surprising direction.

Harding used to have a job doing marine mammal surveys off the West Coast. From Washington to California the survey boat would cruise up and down out in the deep sea, noting down all the marine mammals they saw. Whenever they came across a whale, Heather got to go out on a dinghy to get close to him and then take a picture of his tail to try to identify the animal as part of the vast photo ID database kept on many whale species. "I had the best job onboard, but most of the time nothing was happening. We were slowly sailing up the coast, usually seeing nothing at all. There was a lot of downtime. I was always thinking about bigger questions, about how I could sense my place in a nature that really had a purpose, but I couldn't get any of these scientific types to talk about that stuff with me. They were always writing new computer programs or trying to crack the Rubik's Cube.

"Only in my late thirties did I realize there were all these alternative approaches to science that try to engage our spiritual side as well." Today's she's a science teacher at a Waldorf School, now advocating a kind of Goethean science, which differs from standard biology in that it wants to understand nature by engaging all our senses in observation, following patterns and forms, trusting intuition and interconnection to basic principles of harmony and life.

Goethe received far more acceptance as a man of letters than a man of science, but his approach to understanding the workings of nature certainly has its supporters today. They, like Heather, urge us to engage our full range of senses when trying to make sense of nature: "I want to remain open to both the scientific and mystical approaches to learning about whales," she says, putting a diving mask over her blond-gray hair. "But these scientists believe if we are to work in their paradigm the other way of sensing whales doesn't even exist."

I hear where they're coming from, because a scientist like Jim Darling has to work for years to make even the most tentative conclusions about whales, while someone like Willy believes he can look into the eye of a dying whale and hear the animal's whole life story. It seems so easy.

Rudolf Steiner said that in order to come into real contact with the spiritual world, we have to completely deny it first. That's what he felt our scientific worldview was preparing us for, a reinvigorated sense of touching nature's purpose. In a way that's the same thing mainstream science says about nature: you must throw away all your preconceptions, and only then may you begin to observe, and you keep observing until you have enough data to draw valuable conclusions. Trust only what you have seen and what you have analyzed.

Bennett zooms his boat right beyond the humpbacks' path and stalls it, so the whales might approach us, without us having to chase them. That's not actually getting closer to the whales than the law allows. No crime if the whales come to you. Willy likes to keep it all aboveboard, except when we swim below the surface, waiting for the whales to cross our path.

Snorkeling under, I hear the songs of all the males around us, and then suddenly see three whales emerge into view in the deep blue: a mother, a several-weeks-old baby, and below us, a large male escort. It's the first time I've seen humpbacks from under the water. I'm in the midst of these alien beings, enveloped in a sense of whaleness. It is an amazing moment to be suspended under water as giant cetaceans silently slip by, three by three, on their own determined path.

I don't feel close enough to look them in the eye, but I hear many singers, other invisible whales, some up to five miles away, sounding as if they are swirling inside my head. There's a whole chorus of whales deep down there, swaying and chanting all around. It's the audio equivalent of what Sal Cerchio believes: that it's all a mating lek, with the males gathered around, like a chorus of frogs or a ground where sage grouse boom and perform.

I climb back on board, dry off, get the clarinet out. This time I reel the speaker down, too. I try to play along with this chorus, and I feel like I'm blurring in. Too many whales here, I'm just one more voice. As a player I hear a whole whale orchestra and wonder how I'm supposed to fit in—as just one more male in the mix? Or a soloist, above the fray?

"Listen deeply to the whales, try to hear what they're hearing," advises Heather. "Listen closely and they will talk to you, and tell you what they want. Right now they asked me to tell you that they can't hear you."

I was already worried about that. The underwater speaker does not seem to be churning out nearly enough volume to engage with those 170 dB male humpbacks. But later, when I listen back to this encounter, I do hear some connection between what I play and how the whales respond.

These whales probably hear the clarinet better than I hear them. There are too many males moanin' all at once to be sure what it all means.

"As a naturalist my greatest role is to teach people to see," concludes Harding, "using whatever approaches work best." As a musician I want to teach everyone to listen, whale and human alike. All I can do is follow an aesthetic path, to try and figure out what it would mean to call my human/cetacean music good, or bad. I think Goethe would approve.

Over this year I've been playing with underwater musicians I cannot see, I've begun to dream in whale songs, and sing impossible melodic leaps in my head from low to high as I wake. I'm changing what I play and what I like. Six octave shrieks are starting to become the norm. I'm starting to play whale just by spending hours rocking on a boat with headphones on tuning in to their songs. I try not to copy their sounds but to play things that will fit into the spaces between their motifs. "Gratuitous self-satisfaction," Mark Johnson's Kiwi laugh reminds me I still might be making all of this up. As the whale's song evolves from week to week, even day to day, so does mine.

After several weeks of waiting, the master arrives from Vancouver. Finally I get to go out on the water with Jim Darling, the one man who has spent year after year watching what whales do while they sing. It's early on a clear morning, we're washing off the *Never Satisfied* before we head out. I'm hoping Jim will be able to describe the various parts of the songs as we hear them through the hydrophones sitting there on the boat. I'd like to know which are the latest themes, and which are the most durable ones, sung unchanged for years on end.

"Oh, I can't do that," smiles Jim. Though his face belies a great sense of determination and dedication, he's probably the most humble of all the people I've met on Maui. "Have to print it out in the lab before I can make any sense of the song. I'm really not a very good listener at all."

I ask him if he was surprised by the results of his latest study, which, at fifty-odd printed pages, is the longest, most involved paper on whale songs ever published. "Well, by the time we got to this most recent phase we knew the songs were being sung by males for males. But we thought it would be some kind of dominance-hierarchy thing, not the apparently non-agonistic cooperative behavior we ended up seeing. I hardly trust my own view of this, but I just can't think of anything better." He also comes from the school of cetacean humility.

"But Jim, you've been going out every day you can during breeding season for nearly a decade, can't you trust what you've seen and heard?"

"Yes," he nods, and with a rare moment of bravado: "All those people are still saying the song is for the females because that's what happens in most mammalian displays. We've never seen anything like that and we've really been looking for it. They've just been posing their theories and assuming they're true."

The hypothesis is only part one of the scientific enterprise. Devising experiments to test it is the next step. And when it comes to animals as elusive as whales, carrying out the experiments in a rigorous enough way is expensive, time consuming, and downright tedious. A true scientist can never believe in certainty. He must always be on the lookout for the hole in the theory, and the anomalous result. That's how the next best question will turn up.

"So, Jim, how do you think these males learn the new songs?" I wonder. "From each other, during those moments they silently swim along together after one of them was singing and the other listening?"

"Beats me," he says. "I just can't figure out the mechanism at all. All I can presume is that they hear sounds, they imitate them, and then they remember them." He looks off in the distance, hoping for a spyhop or a fluke to give us an inkling of what direction to head in so we might find a singer. Jim has to record at least one complete song per week to document how the tunes are evolving over the course of a single season. I ask if he's got a good record of all the changes from year to year.

"No, I wish we did. That would just be too much work, and no one wants to pay for something as abstract as that. The money's in management studies, to find out how many whales there are and what the best way is to keep those numbers up. It's difficult to get funding for behavioral stuff, we have to seek out private contributors, people who are just curious about whales and love the same questions we do."

The most astonishing thing about these songs is how they change from year to year, in similar ways even in whales separated by thousands of miles of ocean, in Hawaii and Mexico at the same time, during a breeding season where the individual whales have no contact with each other whatsoever. "Noad's case in Australia proved these songs were learned, not following some kind of internal template like Cerchio believed. But we have no idea how such parallel changes during the breeding season could happen."

"How could the songs in Hawaii and Mexico change in the same way at the exact same time then?"

"I suppose some whales could swim in between."

"Not so likely though, so do you think Cerchio might be right?"

"These guys still keep going with their standard sexual selection explanation, but they haven't been out there getting the data. They just want it to fit the standard model."

"Cerchio says you see enough of the behavior that confirms his theory, but you choose to ignore it."

"Well, I'm not so convinced of my explanation, but it's the best I can come up with. I'm ready to look at his data, but he doesn't have enough of it. He's on to other things."

True, Darling and Nicklin have been watching these singing males and the silent ones who join them year after year. No one else seems to have the patience to do the work and to take the time to raise money to fund it. Like Tutunui the Maori chief, one must be humble and proud at the same time.

Jim is also expert at locating singers and he's got official sanction to speed right up to one of them and position his listening equipment ready. Since he's been doing playback experiments these past few years he has underwater speakers too. "We try to situate ourselves at exactly the same distance that the singers are usually apart, almost one kilometer, and then we broadcast their songs at the correct volume."

"What have you found out so far?" I ask.

"Well, we know we can get them to *stop* singing, but we don't know how to get them to start." He points to starboard. "Look, there's a fluke, a whale just went down. Maybe he'll start singing." We speed right over to where the tail appeared. I take my clarinet out and we toss the hydrophone and speaker overboard.

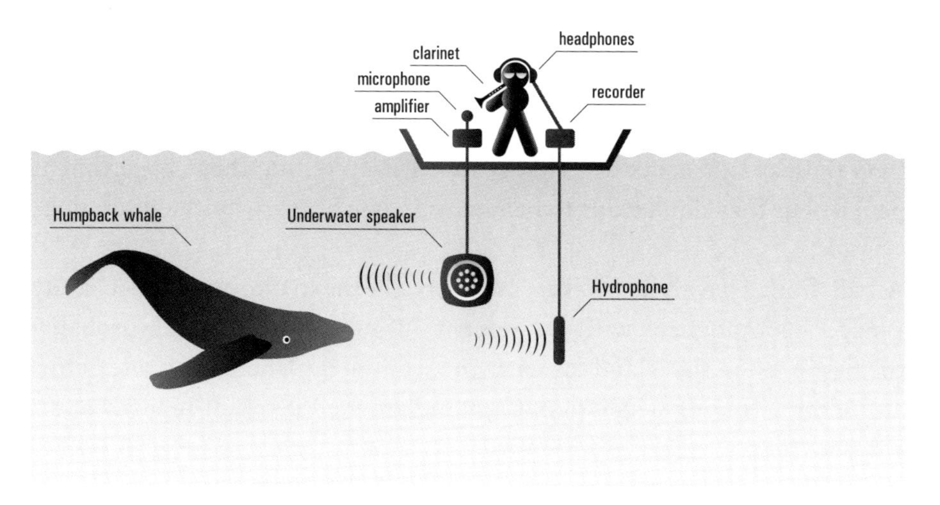


*Fig. 48. How to play clarinet along with a humpback whale.*

Once again there seems to be a brief link between clarinet and whale, mostly heard in the spacing between sounds and the tendency of the whale to jump to a full range from booming low to squeaky high when I play. But for much of the time he ignores me. Do the whales get accustomed to my music and then lose interest? Or are they genuinely in love with novelty—enough so that we could introduce a new sound unit and have it quickly spread through a whole ocean of whales?

"I doubt we could get a permit to do it, but it would be interesting to see if we could specifically introduce a change in the humpback sound," wonders Jim. "My first thought is that it would be really easy. But if it's not easy, and they don't sort of adopt any sound they hear, then it's probably much more complicated than we think. There just aren't that many people looking at this stuff. It's just a huge amount of work, that's why people don't do it. But it's also fairly addictive. You really want to hear what phrase comes next. Every year I give a talk to the locals here about how the song has changed in the current season. Usually it's clear enough, comparing the old and the new, but last year it changed so much, so rapidly, that the whole theory wasn't convincing at all. There is probably much more variation in these songs than most of us are willing to admit."

Just then a whale wing appears in the waves right next to the boat, encrusted in barnacles, and then it swings down. *Splash! Splurf!* I back away, just in time to keep the clarinet dry. "Hmm . . . there's a reaction," smiles Jim. "I wish I really knew what was going on out there," he shakes his head now. "Seems like those males harass the females much more than most of our boats do. That's what I keep telling these government people, but they don't want to believe me," he shrugs, ever the underdog.

A call from Dan Sythe wakes me early the next morning. "Get ready David," he announces with excitement. "I got us all a spot on probably the best boat on the island for what you want to do. There is no one better at finding a singing whale than Captain Samone Yust. You're lucky she's got time for us."

Willy may dowse his way into whale consciousness, but the next day's rogue whale boat has an even more intuitive leader. Captain Samone has the most beautifully decorated of all the boats I went out with on Maui. Her craft is covered with an amazing, airbrushed futuristic painting of cavorting dolphins, stars, and astrological signs. Under the front gap between the two close-set pontoons appears a beautiful golden mermaid. "Oh, the dolphins just love sidling up to the mermaid and rubbing their bodies against her, it's quite a sight to see," says Samone, a tanned and tough-looking woman who looks to have spent years on the whitecaps and swells.

Samone understands at once that I need to find a single singing male, and that we should gently motor along so we will be right on top of him. "Watching for the dive is the key," she assures me. "When we see one flick his flukes in the air and dive down deep, we hope it's a male, and that he'll be down under for at least half an hour, suspended, motionless, fifty feet down holding in place, singing his heart out."

Kent and Dan are along to give technical support, testing out Kent's brand-new stereo hydrophones and Dan's various flash-recording devices. "Tail dive, four o'clock, hundred yards away. We're moving." Samone revs the motor and zooms over there. She turns the engine off, starts to drift. "Lower the 'phones, check it out," and Kent drops down his brand new Cetacean Research hydrophones, one on each side of the boat, to try for a

*Fig. 49. Captain Samone Yust*

stereo effect. "These babies have a totally flat response, all up and down the audio spectrum," smiles the engineer taking a day off from his NASA contract. "Not like that low-frequency biased Aquarian Audio product you've got there yourself."

We have no permission, but we're playing anyway. Scientists are regulated much more closely than rogue whale watchers, so I hope the authorities won't look our way. Kent's copper-tipped hydrophones do get a great sound, swirling around from ear to ear. Not true stereo, because they'd have to be a hundred meters apart to get that, but something more dynamic than what I've yet heard. I carefully lower my ESunPride underwater speaker down its rope. Blow into the mike, test the sound as I listen back to that resonating, underwater clarinet sound I'm starting to get used to. Not quite muffled, more like extra resonances inside an insulated box; more overtones, making it all somewhat bell-like and far away.

*Fig. 50. I finally get through to a humpback whale*

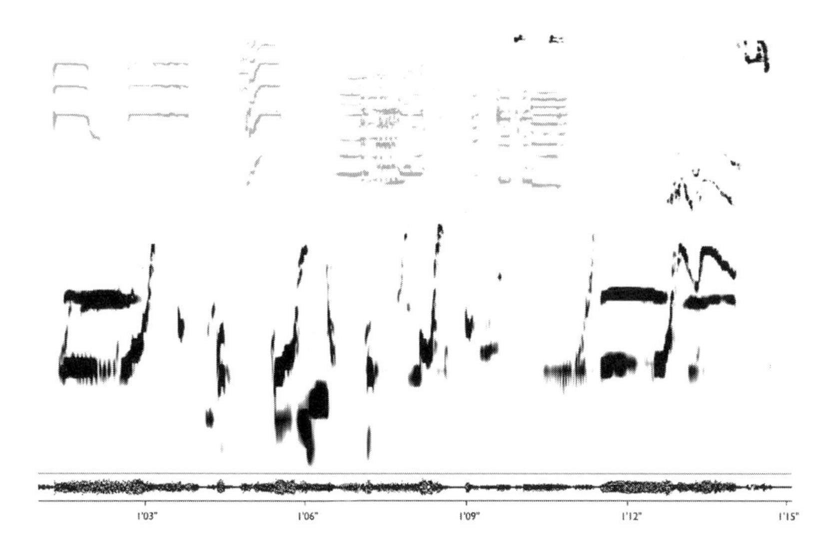


*Fig. 51. 1'0" to 1'15" of the humpback/clarinet duet.*

The whale, on the other hand, sounds amazingly close. Right under the boat. The hull itself is buzzing. How is it *possible* that no one noticed this level of sound before the sixties? I play along for thirty minutes or so, and the whale never stops. Two minutes in, he really seems to get louder in response to the spaces I leave in between my notes. He's alternating with me, not interrupting, like nightingales who compare each other's riffs in the dark. Then I play a high wail, and he *seems* to add a *whoop* to his *bruup*. He's adding resonance to his tones, making them richer, louder. Suddenly he leaps from a real low growl to a super high squeak. The same jump noted by Payne and McVay? Perhaps this one phrase has survived all this time! He sounds like a clarinet at the edge of its abilities. Whale is black and clarinet is gray in figures 51–53.

I've spent enough time with field scientists who deride my claims to make music with animals as wishful thinking, so now I need to assure you that I too have no idea what is going on in this duet. All I know is for five minutes, I felt music. Some of it is harsh and shrieky, but that's what the whales were up to. I entered their wild world of sound. Like Joe Zawinul says, "Sometimes the real music only comes when you have absolutely no idea what is going on."

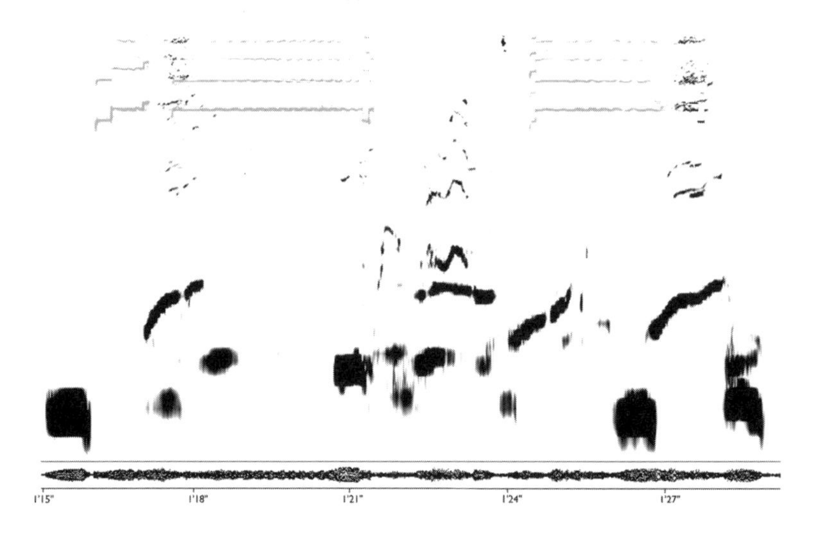


*Fig. 52. 1'15" to 1'28" of the humpback/clarinet duet.*

Finally I have to take a break. Right after I've done so Kent taps me on the shoulder and I take off my headphones. "Stop playing, Dave, I want to check the gain."

"I already did stop."

"What? That whale has started to sound just like you."

When I get home from Hawaii I hunker down in the studio, replaying the experience over and over again. It's the middle of winter, cold and gray. I listen to my tapes, I equalize them, compress them to bring all the sound levels up so the whale and clarinet are closer to each other in volume. I listen to hours of recording from all the different boat trips where not much is happening, then home in on the best several minutes where alien music seems to appear. Then I play this section for everyone I know, and I ask them, "Is that whale listening to me? Is he responding?"

Far away from science, I know. But it's a music like none I've ever heard before. Most who've heard it thus far also hear some kind of connection between man and whale. The best recording of all comes from Samone's boat. Listening to this one several-minute interchange, it appears that the whale and I do get through to each other.

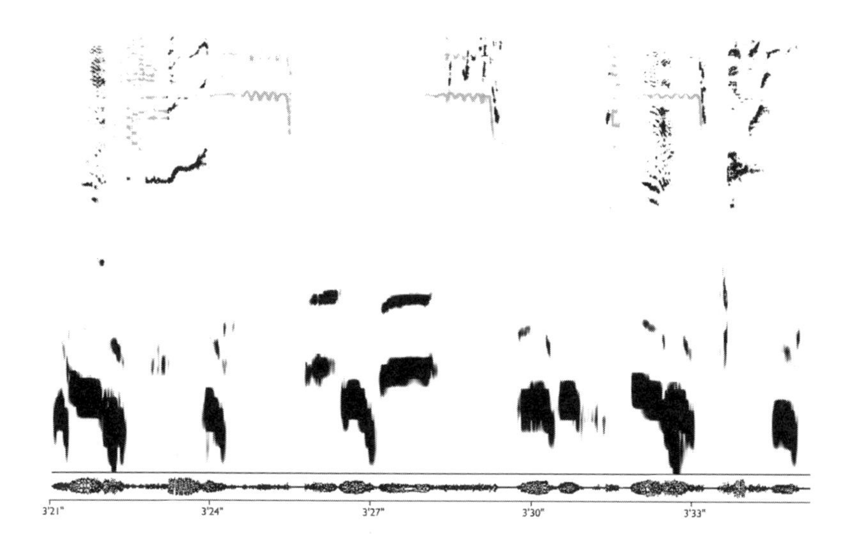


*Fig. 53. 3'21" to 3'34" of the humpback/clarinet duet, the wailing climax.*

As the duet continues, I hear that the whale makes use of the spaces between my phrases, filling them in with patterns from his repertoire that at times seem to echo the pitches and timbre that I am playing. He leaps from his low beats and inserts his shrieks just when space allows:

After several minutes I'm flowing deeper into the style, ready to shriek wildly into these klezmeresque vibratos, evenly spaced, like the whale aesthetic that I'm hearing. I suddenly remember something I once heard about Sidney Bechet, practicing his soprano sax in a Paris apartment in the 1950s. He'd play scales and arpeggios for hours, but at the end of each session, he would conclude with a series of howling animal noises. A neighbor in the building once asked him about this, and he paused a moment and just said, "Sometimes I think what we call music is not the real music."

In a way it is no surprise that this whale is able to match my unfamiliar sounds so quickly. If he couldn't do this, there would be no way humpback males could learn new songs so quickly, changing their group song rapidly over the course of one single season. A new whale could change the song, and so might a clarinet. This is a hypothesis I have just begun to test. Listening back later and printing it out, I look at the page, and the pattern tells me there may be some sense in it after all.

I'm screaming, wailing, screeching too, trying to pack as much emotion as I can into one moment when a whale wants to listen. It roars through the seas, only for a few minutes, this music neither human nor humpback could create alone. Yet each musician makes space for the other, our duet an overlap of themes from different worlds, human and cetacean. Undersea and above, it is music with no beginning or end. We pause only to back away from the possible sounds that still remain. Then we stop paying attention to one another, back in our separate worlds. Neither human nor whale has forgotten the song we made together.

There are more humpbacks in winter Hawaiian waters than ever before. Human support for whales has grown nearly universal, and the whaling moratorium of 1986 has truly worked, becoming a conservation success story. Of course there are still countries like Japan and Norway who want to defy international law and bring back a full-scale whaling industry. But the world is against them. Like those few remaining dissenters against the rising reality of global climate change, they will eventually give in. In August 2007 Iceland announced it was suspending its commercial whale hunt for lack of demand for the product. Most people want to save our planet, and most of us want to save the whales.

It was the song that did it, that got us to care. Maybe you heard the record years ago, maybe you've heard whale music in the movies or at the tail end of a pop song. Maybe you recorded the most beautiful song yourself on a lonely Valentine's Day many years ago. However you find it, once you hear the song of a whale your life will be changed.

It's quick to escape the winter down to Puerto Rico, and there's an early flight to Tortola out of San Juan. I get there at dawn and drive the impossible paved-over goat path back down to the beach. There is Paul Knapp, making breakfast on a little stove outside his tent. From the looks of it, he might have been sitting there all year.

"So David, glad to see you're back." We sit down by the fire, boil up some water for tea. Nothing has changed on this beach in one year; that's good to see.

I tell Paul the legends linking human and orca from Puget Sound to the Johnstone Strait, how every family of whales up there is known by its calls. How no one is really sure what all those dolphin whistles mean. How we know even less about the vast, alien music of beluga whales, whose sleek, white forms surface in Arctic seas, where belugas do not believe in tears. How tiny clicks organize pods of sperm whales across the world's oceans, how we're still not sure if whales really listen in the natural internet of the deep sound channel, even though the military is booming sounds down there—something the law insists they stop. How among all those humpback whale scientists and would-be whale whisperers, I never met one who was as much a connoisseur of the songs themselves as Paul Knapp, the one man who goes out to find whale songs every day, not to figure them out, but only to listen.

"Did you manage to do it, did you ever get through to a humpback whale?" He hands me a cup of hot tea. The wind rustles in from the ocean through the palms, blending its sound with the bleats of forest frogs.

I gaze out toward the mouth of the bay, remembering all I've heard.

"The more music I make with whales, the less I know. Maybe he listened to me, Paul, or maybe not. But have I got a song for you. . . ."

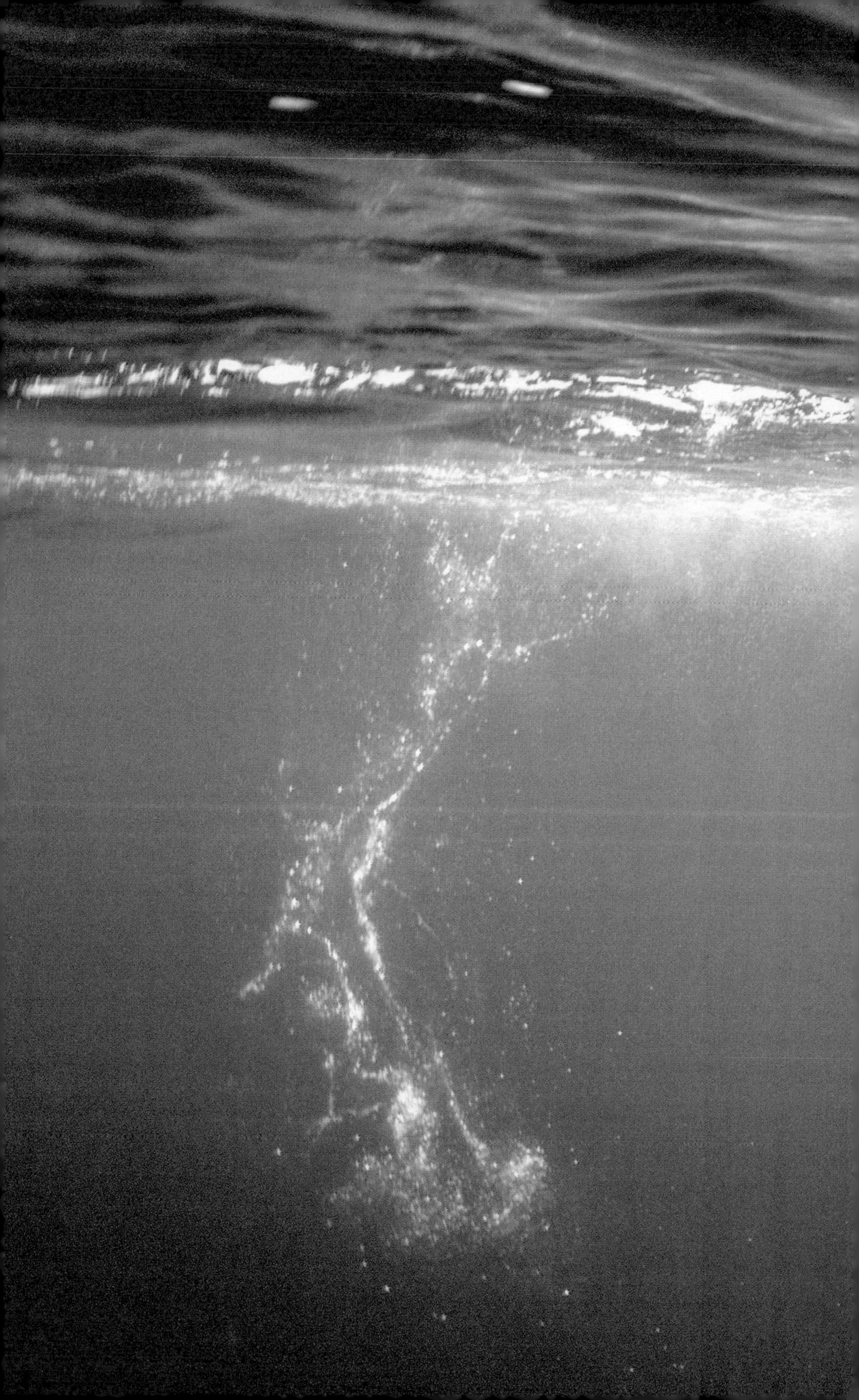

# 10

## WHALE SONG INSIDE US

### Notes from the Future

> What is it about whale song that feels, still, so futuristic? Is it that whale song holds the post-apocalyptic bona fides of a language that has survived the eradication of so many of its speakers, worldwide? Or do we imagine that the songs of whales are tantamount to the voice of the environment without us? A misanthropic fantasia; reversing stories of the starry interior of the whale, in which Jonah crouched, and turned, again, to Eden.
>
> —Rebecca Giggs, *Fathoms*

Well, that was a pretty good ending. A circular tale really, back to the Caribbean beach where the whole journey began. Nice job Dave. It was 2008. I still feel this is my best book.

Of course it wasn't really the end, but more like the beginning. The start of a love for whale songs and a desire to join in with them, to share them with musicians and audiences alike, to work hard to convince people that these bleeps, bloops, clicks, and shrieks are closer to music than to language or noise.

The first edition of this book came out in the fall of 2008. Its editor quit. There was a financial meltdown. People were obsessed with other things. *Thousand Mile Song* didn't exactly make a splash. Still, a few people found it. They wanted to hear more. I continued to travel the world to find whales, and people who believed we could get closer to whales, with music being one of the most profound ways.

Now it's fourteen years later, and the world above water has changed. The constant stream of information has increased a hundredfold. We are awash in data and believe it all can save us if we can only convince our machines to listen and learn to it all. Music has gotten noisier, weirder, even more open to unusual beats and squeaks. Our human music has only grown closer to the sounds of whales.

So can we now get it? Can we get inside the mind and art of these whales? We're still not there yet, but it has sure been fun to keep trying.

Perhaps the most important change over the last decades is the growing acceptance of the idea that seemed quite radical when Hal Whitehead first proposed it: that whales with their social learning and diverse sounds and lifestyles can truly be said to possess culture. As you may recall, when this idea was first published in *Behavioural and Brain Sciences* it was followed by literally fifty pages of commentary by critics from many disciplines, all to muse on whether a group of animals living socially underwater, far from the methods of human habitation, could be said to have anything so erudite-sounding as "culture."

In a recent review of several books on animal culture in the *New York Review of Books*, philosopher Martha Nussbaum champions the "Capabilities Approach," which values animals and their situation on the basis of this question: "What is this creature actually able to do and to be?" She takes this as a starting point for a new theory of justice for more-than-human animals, so we can now value their physical, emotional, and cultural needs with more gravity and weight. Such a new understanding of what it means to treat other species fairly could lead to whole new ways of interacting with the myriad other cultural species with which we share this planet.

The idea of whale culture is so accepted today that it is the main theme of Brian Skerry's excellent *Secrets of Whales* series for National Geographic (now part of Disney!) TV that came out in 2021. Culture is illustrated so graphically after Skerry shows an orca slide onto the beach and grab a baby seal, which she then drags surreptitiously into the waves to become her next meal. Just a few minutes later there is footage of seals and killer whales fishing together off ice floes in Antarctica—the same two species, but living differently together. They have learned, not inherited, these different ways of life. Foes on one continent, friends on another. Now generally recognized to be cultural animals, this is the kind of behavior we have come to expect.

Speaking of *National Geographic*, I wrote in chapter two how important the magazine was to the music industry when it made the greatest single pressing of any record, ten million copies of a sound page of a few humpback whale songs, back in 1979. Music distribution has become something completely different in the four decades since that time, but I still thought *Nat Geo* should do something about the new science of whale music and how to visualize it. In fact, I have been nagging them about this since *Thousand Mile Song* came out in 2008. They simply ignored me, until last year. You'll find out below what happened.

Why did I change the title of this book to *Whale Music*? Another thing that's happened in the fifteen years since the first edition of this book is how people find what to read and listen to—getting just the right search terms so the algorithms find you. Though I still love the image and reality of the thousand mile song that resonates widely in the deep sound channel, the beautiful image didn't help readers find my book. I once told Vanessa Carlton if you type in "thousand mile song" you either get her hit tune or my book, and she did laugh. She even went on a Cape Farewell expedition to the Arctic to help fight climate change through music.

*Bug Music. Whale Music. Pond Music.* There you get some of the subjects I'm obsessed about. I just keep telling the same stories with different creatures, distinct ways of attaching the human to the natural. Still, whale music still touches me the deepest and moves me the farthest. I still want to know why, and I still want to convince people that some of these sounds are closer to music than to language.

How to get people to take whale music more seriously? I felt the sonogram, that essential tool of the animal sound scientist, needed to be improved, to be made more beautiful, more accessible.

I knew data visualization had grown into a well-traveled obsession among the Silicon Valley set. So I kept asking people I knew who might help me come up with more exciting, more beautiful imagery of the frequency of sound vs. time.

One name kept coming up: Michael Deal. Designer of the Pinterest logo, longtime Google employee, later design director for Splice and for Vice. Eventually I met Michael and told him to let loose on whale song visualization without reading too much about how it had been done before.

First, we wanted to come up with the simplest, smallest list of motifs that make up the full humpback song. Here we choose one famous song recorded by Paul Knapp because it is so clear, complete, and comprehensible to sonogram technology and to human ear alike. Today, sonograms can easily be spewed out instantly by laptops or even smartphones using software like Amadeus, Raven, or Sound Analysis Pro. Yet the sonogram is still a somewhat inscrutable scientific printout. We wanted something simpler, clearer, easier to grasp. So we decided color was the key. Taking the original Payne and McVay diagram as our impetus we broke down the song into its most basic set of units, indicated on the original cover in distinct colors.

This song ends up having about ten distinct units, which are assembled into motifs and phrases, as Payne and McVay's original hierarchical diagram indicated. That song was recorded off Bermuda in 1967.

We apply the same technique to a particularly fine humpback whale song recorded by Paul Knapp on Valentine's Day 1992, twenty-four years later off the shores of Tortola. This song is composed of nine distinct units. Here we map them against sonogram printouts of the same sounds to demonstrate the "glyphs" that Michael Deal invented to stylize the sonograms into shapes that might be suggestive and not over-detailed. The top row contains individual examples of each unit. These colored forms were created by tracing the "averaged" shapes that resulted from overlaying the many occurrences of the same unit across Knapp's recording. All right, these shapes might be easier for human viewers to perceive than the raw sonogram, just as McVay and Payne's simple tracings made more sense than the primitive sonograms possible to make in 1971. But we needed to clarify that these sounds also contain pitches and overtones that are perceivable to the human, and also the cetacean, ear. So we decided to combine our shapes with musical notation.

Each solid horizontal line in the staves of standard musical notation represents a different number of Hertz along a logarithmic frequency scale. These whale song sonograms share the same logarithmic scale, conveniently allowing us to combine the two visualization contexts. Because standard musical notation is, in essence, made of timelines of note symbols plotted against a vertical axis of pitch frequencies, we can match the whale sounds to their corresponding frequencies on the musical staves. Hopefully this gives the whale sound shapes a more familiar context.

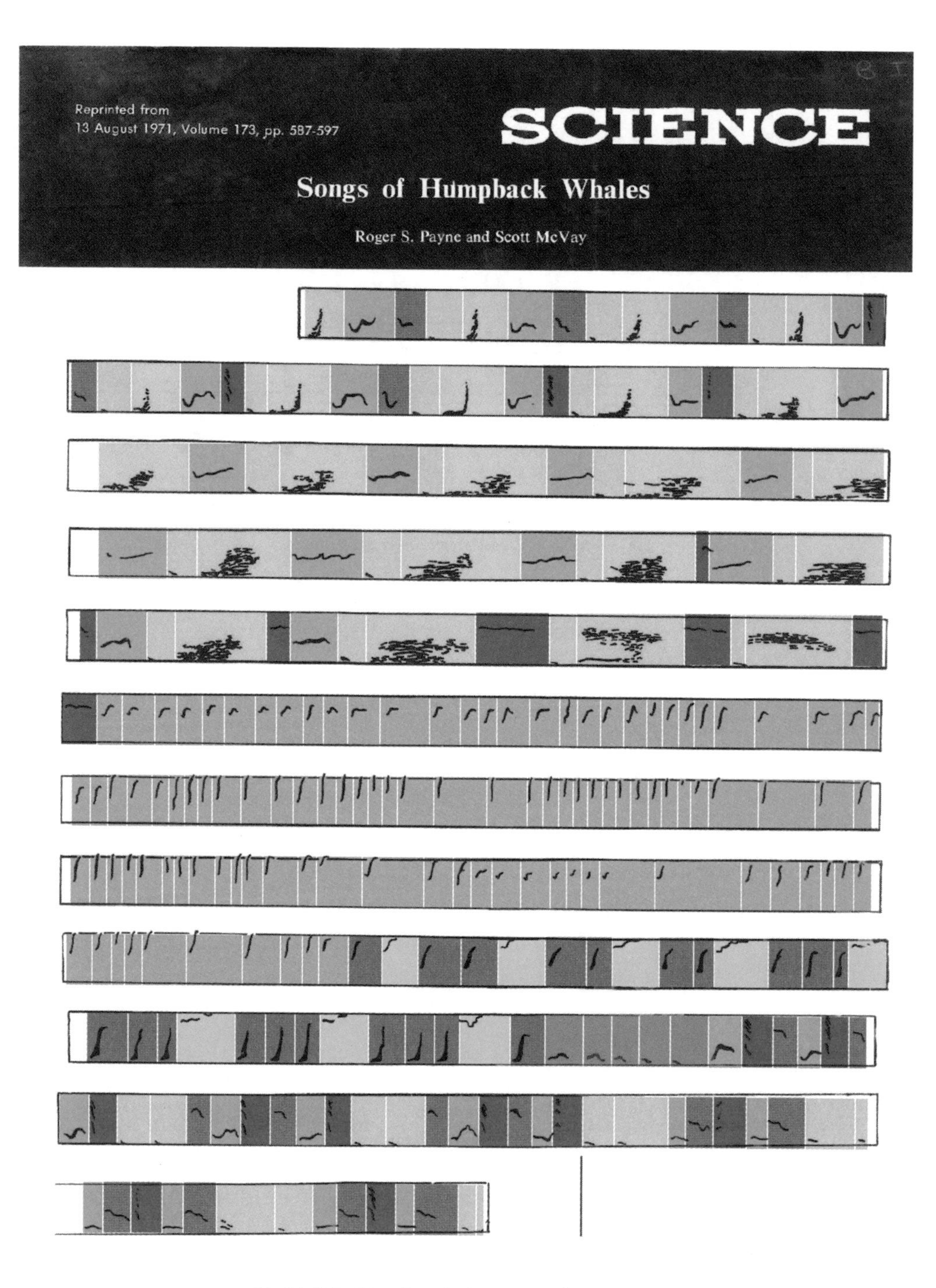


*Fig. 54. Colorizing the units in Payne and McVay 1971.*

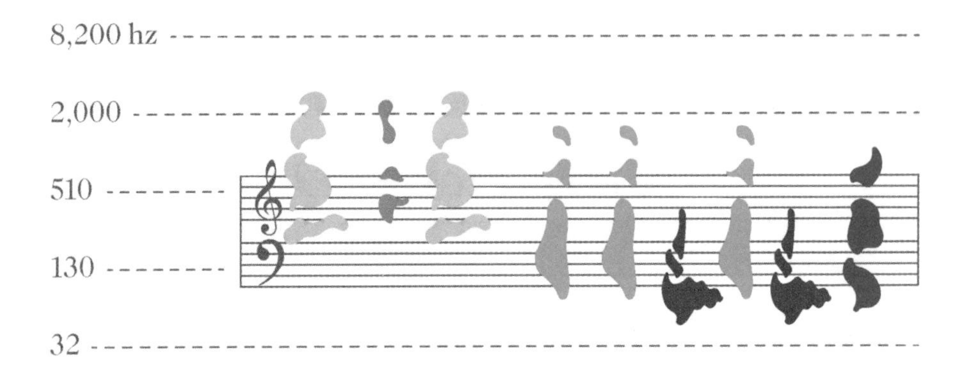


*Fig. 55. Michael Deal's glyphs on musical staves.*

The next step is to map out the entire twenty-minute whale recording using these techniques, blending the nine basic glyph units with musical staff notation into a perceptible whale music score. The beginning of Knapp's recording appears in figure 56.

We wanted our new visualization of the humpback whale song to change human perception of this unique natural phenomenon, and we hope that this hybrid sonogram–ideogrammatic–musical notation helps the human listener make sense of this song as we listen to it.

For a sequence of sounds to be considered music it should have form, pattern, and a sense of rhythms and pitches that can be identified as falling within the realm of possible music. That is why we placed our motifs on musical staves, to show how they might be overlayed with human music. But each whale sound is not a clear pitch on a human-defined scale, or a precise measured rhythm. That is why we developed unique glyphs based on the sonograms that machines can spew out. Adding the color helps identify sameness, repetition, and difference, the statistical elements of musical order.

The periodicity of pattern and change in humpback whale songs is a bit extended for most humans to make sense of as we hear it. That is another reason such visual notation can be helpful in aiding the perception of structure in sonic phenomena that proceed at a pace too slow for humans to easily know. Visually we look at the "score" and grasp the structure at once, hopefully with more clarity and subtlety than in previous notational attempts at analysis of humpback whale songs.

As whale song entered the realm of popular artistic inspiration, so it began to be taken more seriously by science. Over the past half century we've learned much more about it: Only the male whales sing, so it is generally assumed by scientists that they sing to attract the attention of female whales. However, in the five decades people have been studying this phenomenon, we have *never* seen a female whale show any visible interest in the song. So maybe it's all about something else...

In any one ocean, all the humpbacks sing roughly the same song. But the song does not remain the same. As an ocean-wide population, the whales change the song, all together, gradually evolving new phrases and patterns from week to week, month to month, and year to year. Over the decades we can trace the gradual change of the phrases.

If they all sing the same song, why do they need to change it? Such change is rare in the animal world; birds don't do anything like this. Some believe the change is of interest for the sake of change alone, like our constant need for a new hit song. Maybe the whales, like us, just get bored with the old tunes... If so, one wonders why other musical animal species don't try the same trick.

So just how different from the glory days is the whale song of today? Roger Payne has repeatedly said that the whale song of the sixties is far more beautiful and deep than anything the whales are singing these days, but people usually feel that way about the pop music of their youth.

It is a desire to go beyond memory and into the sounds of reality that has led me to release *New Songs of the Humpback Whale*, which aimed to gather the best recent recordings of scientists and whale listeners the globe over to give all of us a chance to assess what has happened to whale song over the past few decades. (Details of this release, along with all the other albums I've put out since 2008 based on whale songs, appear in the

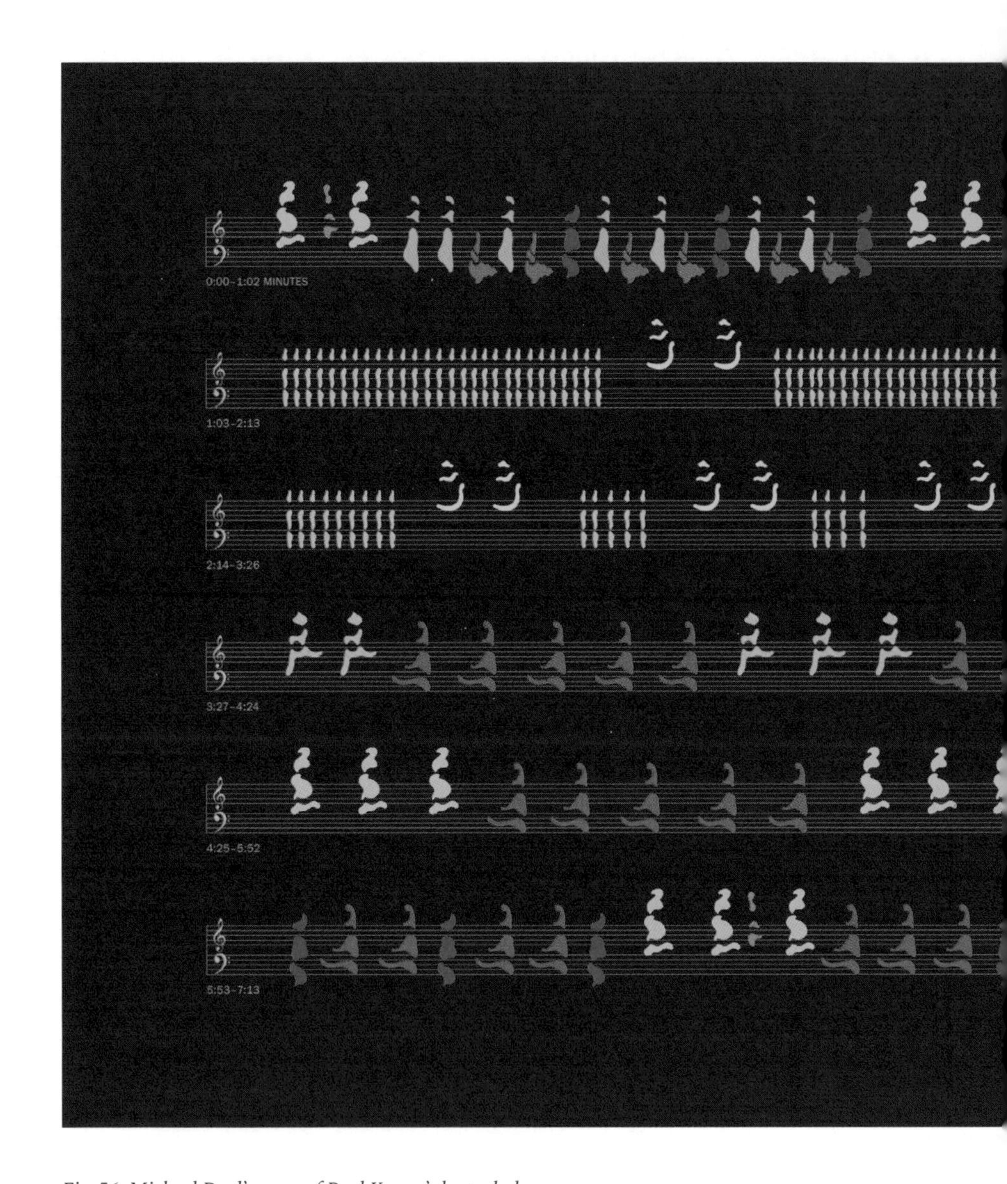


*Fig. 56. Michael Deal's score of Paul Knapp's best whale song*

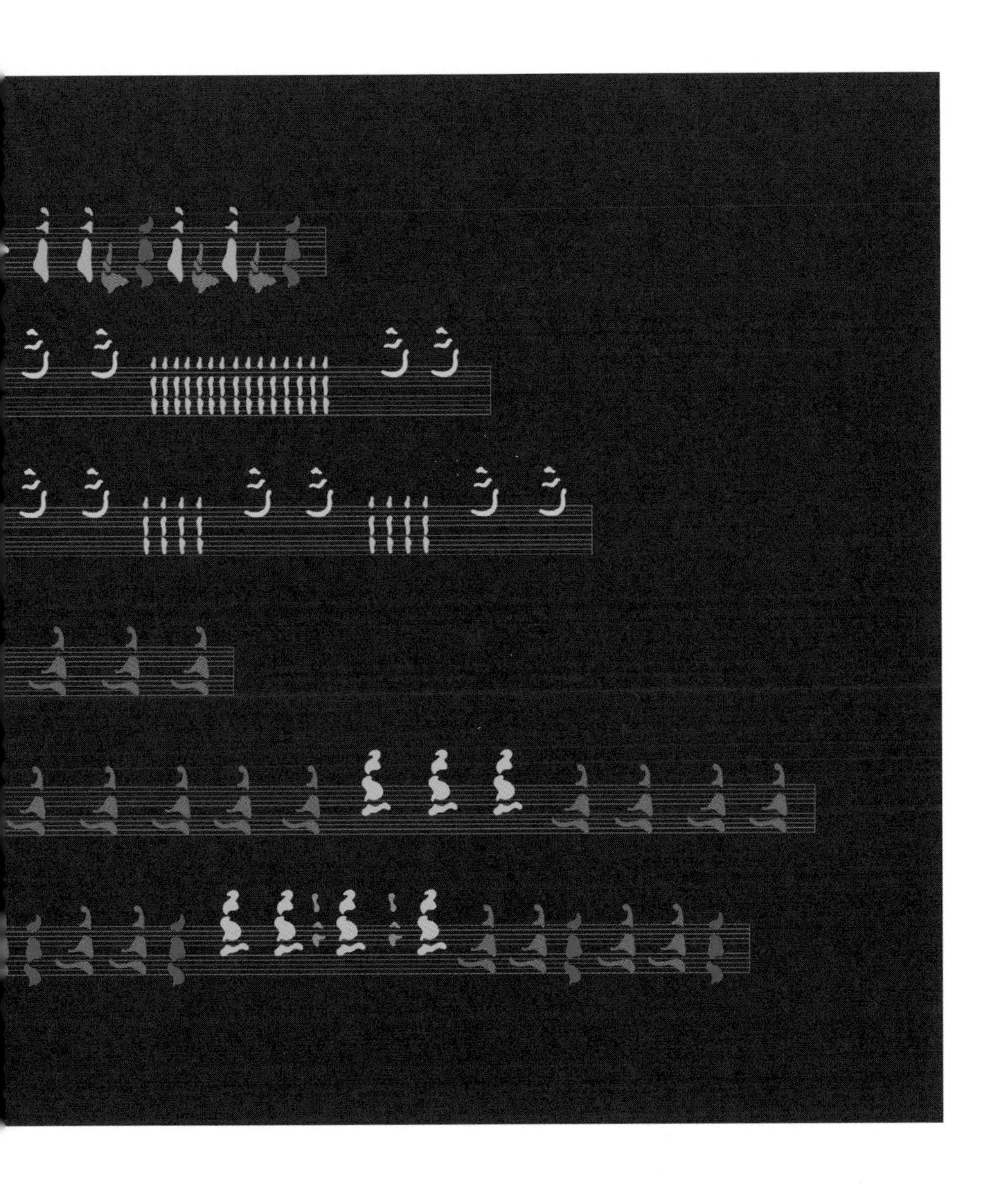

"Liner Notes" section of this book that follows this final chapter.) One can hear gradual changes in humpback song from year to year, with some phrases lengthening, others shortening, others disappearing altogether as new variations appear. The change can be heard month to month, even week to week, within a single season. So Roger Payne is right: Today's whale songs should sound quite different from what he fondly remembers from the sixties. But is their musical culture going downhill? Or is it on the rise? Or just changing? Can we humans even *tell* the difference?

One of the biggest changes since the 1950s, when Frank Watlington started recording these humpback songs, is how much easier it is to record them today. Reasonably inexpensive hydrophones from Cetacean Research or Aquarian Audio can be plugged into inexpensive digital recorders or even, with an interface, smartphones, and whale song recording is accessible to many more people today. Do these more portable and more stealth technologies produce different sounds? I think so. The recordings on *New Songs of the Humpback Whale*, from whale scientists as well as whale enthusiasts, sound crisper, more angular and noise-based, less pretty and no longer drenched in reverb and delay, than Payne's tunes from the sixties.

Such complex animal songs are actually quite rare in nature, but at such levels of extreme beauty, there are strange parallels. Speed up a humpback whale song, and it sounds surprisingly like the song of a thrush nightingale, with a similar balance between rhythms, jumps, and long clear tones. Both of these animals are "outliers," with unexpectedly beautiful and complicated songs. In neither case can we accurately explain why such a song needed to evolve so extensively. But aesthetically, there are definite parallels. Does such a parallel mean anything? Perhaps there are basic principles at the root of what different species understand to be beautiful. Evolution, as Charles Darwin knew well, is much more than survival of the fittest with natural selection, but it also includes survival of the beautiful through sexual selection, which is supposed to explain why whale songs are so long and moving, even though we have yet to see a female whale show any reaction to it. Of course why would they want us to see something so private anyway...?

Though humpback whale song did not evolve for humans to appreciate, it may be no accident that we do. The beautiful has evolved in the same world in which we have evolved, and this may be one reason we are always drawn to nature so much.

The video version of Mike's color-coded sonogram proved to be especially popular. Watch the glyphs cross the line, and sing along with a music from another species. [https://www.youtube.com/watch?v=NXaxWKzTaRc] While showing this once at the Marble House arts residency in Vermont, I casually mentioned that when sonograms were invented in the years right after World War II, it was thought that these "objective" visual representations of sound might help the Deaf learn to speak, and I said that application hadn't really panned out.

After the talk an artist at the residency, Maria Batlle came up and said, "You know, I teach music to the Deaf in the Dominican Republic, and I know we have plenty of humpback whales there. When my kids see these cool color sonograms, they will be singing along with whales in no time."

"Hmm," I said, "we're going to have to try this out."

When I first heard about Maria Batlle's MuseSeek Project I knew that it would be possible to get deaf children to experience and join in with humpback whale songs because the structure of the songs can be seen in very clear visualizations shown above and the sounds themselves could be felt by those who cannot hear by wearing Woojer SubPacs, devices for DJs to let them feel the bass, and just cranking up the lower frequencies as loud as possible.

The three kids we worked with were very different. Joel Fernández Rodríguez was born and raised in Samaná. There is no school for the Deaf in Samaná, and Joel has never gone to school. He does not know how to communicate with his family or friends other than with facial expressions. Luis Alberto Cancú was born in Samaná, raised in La Romana. Luis goes to a special school for the Deaf, and now he is the first deaf person to be admitted to the Altos de Chavón School of Design and also the first to perform a primary role with the National Ballet. David Esteban was born and raised in Santo Domingo. He receives sounds from his cochlear implant, and during this activity he took off the external part because he wanted to *just* feel the sounds with the SubPac. His parents are activists for the Deaf in the Dominican Republic.

Our journeys to jam with whales took place in the bay surrounding Samaná, a famous destination for whale watching in the Dominican Republic. Famous it may be, but no local kids like Joel or Luis would even *think* about going out whale watching—that's just for rich white people. They loved the opportunity to get out there and see and even hear these whales by feeling their sounds and studying their structure.

All of them, despite their range of education and experience, quickly figured out how to sing and play along with the songs of whales coming up from the hydrophone, and I have never seen a group so filled with surprise and joy at being able to join in with these great subaquatic melodies.

This certainly wasn't a direction I expected this work to go. If anything, we proved that sonograms could help the Deaf learn to sing and get one step closer to the undersea world of whales. And doing this in the Dominican Republic was a bit of an about-face in the whale-watching business there. Usually the whale people are rich white tourists from the North like me. But we were taking out children of color from local villages, and from one of the best schools for the Deaf in the Caribbean. Listen to the whales beyond your ears. Bring the people to the sound, those formerly excluded. The world is changing.

This whole experience, culminating with a hundred deaf kids feeling the giant whale songs reverberating through subwoofers that shook the building of the Centro Leon in Santiago while watching the sonogram movie stream by, reminded me just how important the visualization of complex sounds can be in helping us comprehend them. By delving into the diagrams, the music of these aliens beneath the sea comes ever closer to our understanding.

Over the years following the WhaleMuseSeek experience, people got more and more accustomed to the appearance of large data sets as pictures through the exponential availability of information in our algorithmic world. People had been recording humpback whale songs since the 1950s now, for more than sixty years. What could be done with all this information? Do we have even close to the patience to listen to it all?

First we can look at it, then we can sift through what we see, and make sense of the layers of structure. First everyone needs to get

Fig. 57. *Hands of the deaf feeling the clarinet sound. A still from* WhaleMuseSeek, *a film by Maria Batlle*, (2015)

access to the same plethora of data. Here's where Google got into the act. There's a division of that company called Google Creative Labs, the people responsible for a series of projects called Experiments with Google. What a great job! A handful of the most creative and smartest software engineers get to collaborate on cool stuff, entirely for the fun of it. (Just remember: Don't be evil!) https://patternradio.withgoogle.com/ presents 8000 hours of pristine Hawaiian humpback whale song, recorded by NOAA from 2014 to 2015, entirely for your enjoyment and investigation, however you see fit. Google asked a handful of people to offer personalized tours through the data, including oceanographer Ann Allen, biologist Chris Clark, composers Annie Lewandowski and myself, and a class of seventh graders. Eight years later, maverick whale song researcher Eduardo Mercado is asking citizen scientists to help him parse this data, inviting everyone to participate in his continued attempt to find evidence that humpback whale songs are a new kind of poorly understood sonar. It doesn't faze him one bit that most scientists completely disagree with him. Right now he just needs *you* to help him sift through all that data.

Meanwhile other researchers think it's not your fingers that should do the walking but the machines themselves. Interspecies Internet began with a conference in 2019 at MIT spearheaded by Neil Gershenfeld from the Media Lab, featuring some of the most important names in animal communication, from Roger Payne and Sue Savage-Rumbaugh to Aza Raskin and Britt Selvitelle from the Earth Species Project. Funding for the initiative comes from Google and the Jeremy Collier Foundation, and it is all just beginning.

The basic idea is that we have been going about automated translation all wrong. We think machines need to learn the fundamentals of grammar and the rules of language from one tongue to another and be taught these structures by the humans who program them. But we have bigger data sets than ever before. Why can't we let the machines learn on their own, through trial and error, through crunching of the data far faster than humans could ever do? What if, not data processing, can computers most excel at?

Letting machines learn how to translate on their own after comparing ten thousand hours of French and German led to mutual comprehension at an astonishing 75% accuracy, without the machines having been taught any specific rules of the two languages by us in advance. Why not, thought Raskin, let the machines loose comparing bat and dolphin languages? Whale vs. nightingale? Finally the umwelt of other species might be genuinely accessible.

Well, I mentioned to Raskin at that initial MIT conference that because people speaking French and German are basically doing the same things, I can see how that translation might actually work. But humans and whales are living completely different lives. What would translation even mean between such different lifeworlds? "That," he responded, "is a fair critique."

But naysayers like me haven't stopped these bold dreamers from going forward. The most promising outgrowth of this vision is Project CETI, which aims to use the machine learning approach on the many sounds made by sperm whales in a study area in the Caribbean. As mentioned in chapter five, we know the clicks made by these animals combine cultural information with sonar information. What will the machines make of this? They hope to have the first results by 2026.

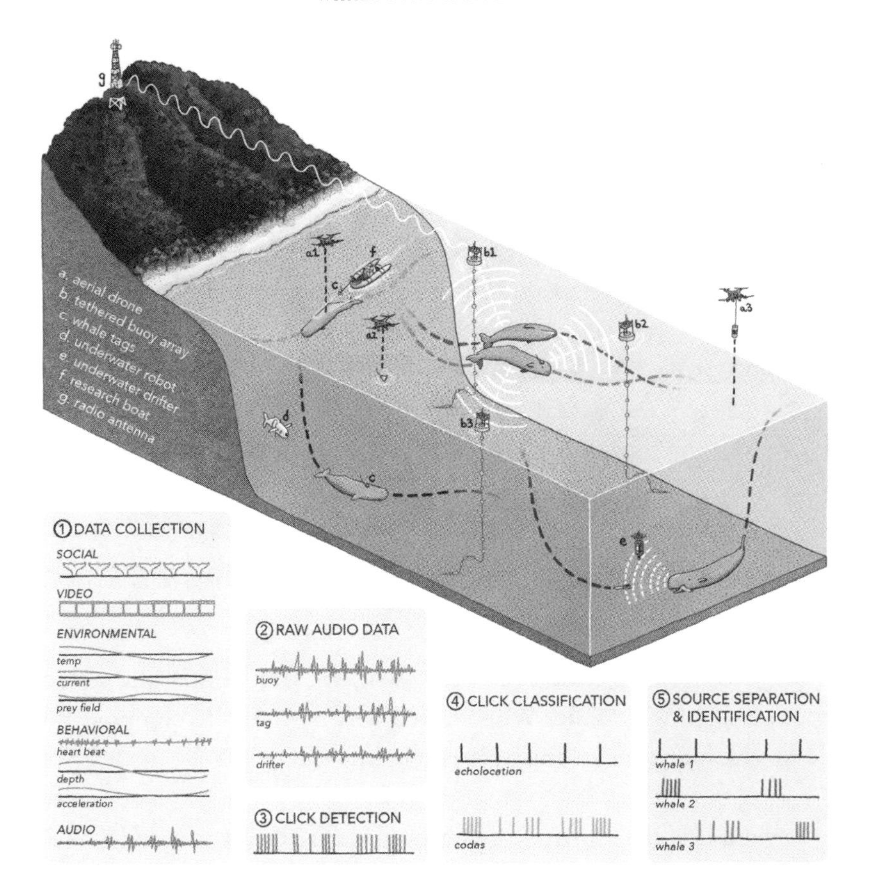


*Fig. 58. How Project CETI plans to use machine learning to classify thousands of hours of sperm whale clicks.*

As exciting as the possibilities of human/whale machine translation might be, based on some kind of phenomenology where the "whale-in-itself" presents his sounds to our devices, who sift through them with no advance preconceptions, I am more excited by the increasing number of humans aiming to play music with whales, either live or in the studio. Just as I was inspired by the interspecies music of jazz saxophonist Paul Winter, who is still going strong in his eighties today, there are new generations interacting with whale music, more and more people with each passing year.

Aline Penitot is playing her bassoon with humpbacks in Reunion. Abigail Sanders is composing original works in Berlin for solo French horn and humpback whale, learning in detail from each phrase of the whale's famously emotional song:

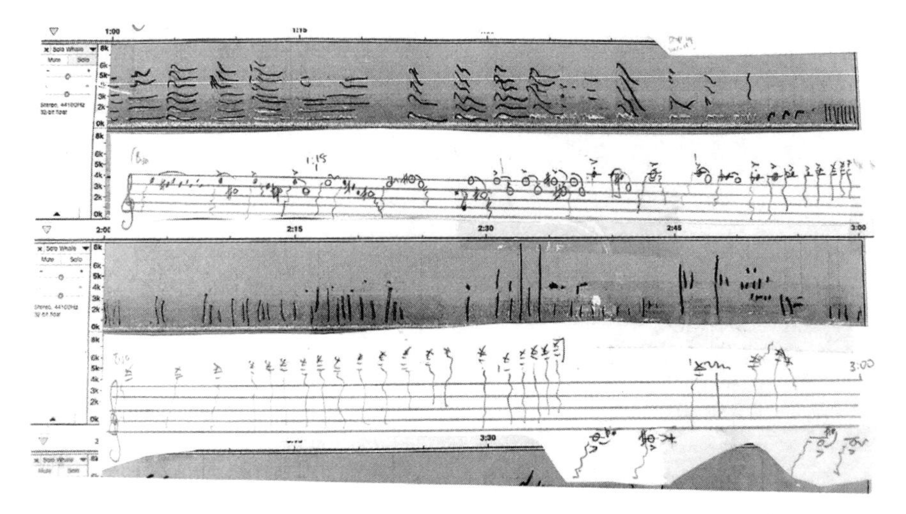

*Fig. 59: Excerpt from the score for Abigail Sanders, "Solo Humpback" (2022) for french horn and recorded whale.*

Alex South is working on the world's first Ph.D. in cetacean zoomusicology in Scotland. When asked why he wants to compose music based on the humpback song, this is what he said:

> For me it is the shifting themes and transitions that can be the most moving and aesthetically satisfying parts of humpback song, and they seem to be easy to dialogue with whether improvising or composing. These features seem to bring human and humpback musicalities into contact, but are only one side of the story. I am equally interested in the differences. Unlike the rhythms in much of human music, humpback phrase rhythms do not appear to be organized around either a steady beat or common fast pulse. Another difference is that humpback whales singing within earshot of each other do not appear to synchronize their songs. Whatever the reasons behind them, these kinds of differences are exciting to me as a musician because they offer me new ways of putting sounds together that I hope result in pieces that possess something 'humpback' about them.

Here's the beginning of one of South's scores, yet another work for bass clarinet and recorded whale:

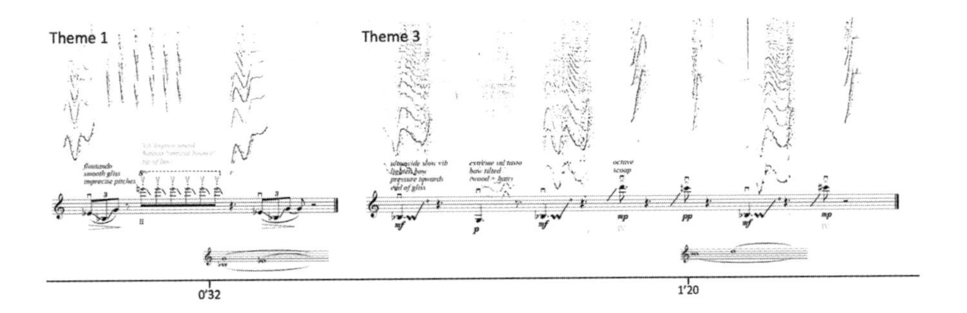


*Fig. 60: Beginning of the score for Alex South, "Whale Bow Echo" (2021).*

Perhaps the most elaborate new musical work made in tandem with whale sounds is the installation *Siren* by Annie Lewandowski, created in collaboration with Katy Payne, Amy Rubin and Kyle McDonald.

What I like about *Siren* is its many-layered approach. Lewandowski spent the season of 2019 out with Katy Payne listening to and charting humpback whale songs in Hawaii. Yes, the great whale researcher Katy Payne is still out here listening to whales, diagramming them and studying. She is just less interested in publishing her work prematurely like so many other contemporary researchers. She takes the time. And Annie was out with her.

After weeks of recording and then months of poring over the recordings identifying themes, Lewandowski hooked up with Kyle McDonald, one of the masterminds of Google's Pattern Radio project. Their goal this time was different: to create a multisensory artwork that conveys the beauty and mystery of the humpback song.

Lewandowski gave McDonald her analysis of five clear themes by annotating, or tagging, individual humpback units, or notes. He wrote a program to listen to many hours of recorded humpback songs to identify these specific themes, paying particular attention to how similar they were each time they were sung. From that he created "similarity maps" to categorize each theme in a less linear way than we are used to seeing with sonograms. The more concentrated the cluster, the more similar

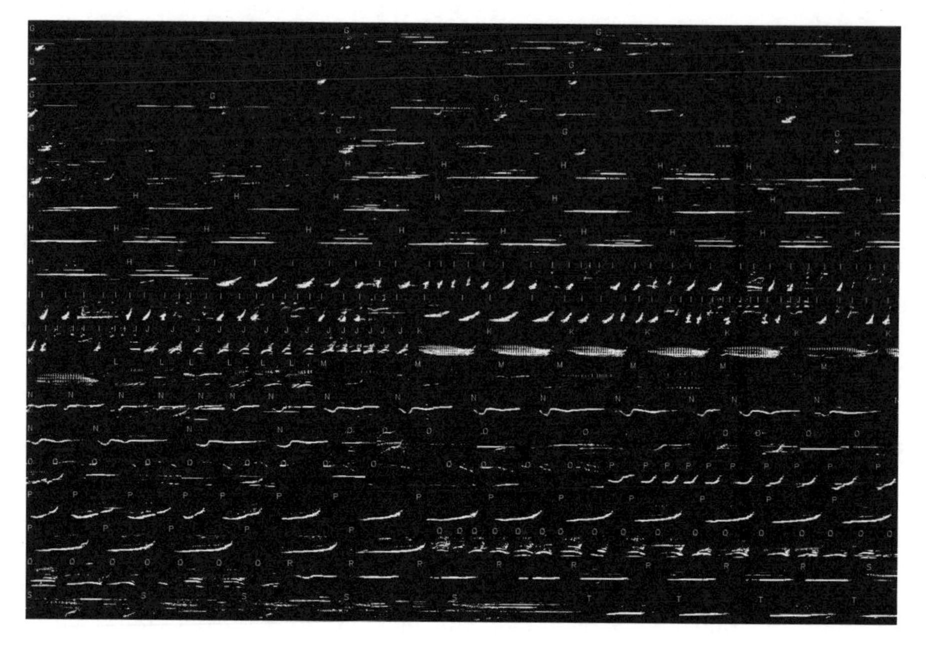

*Fig. 61: Machine-sorted units of humpback whale song by Kyle McDonald.*

each theme is when it is sung. The "supervised" versions, where the software has been fed information about each song unit noticed by Annie the human observer, are much clearer than the ones where machine learning tries to identify song units alone. A lesson, perhaps, for those pure machine learning dreamers out there!

In addition to a composition made out of the songs from five distinct humpback song sessions, *Siren* includes a sculpture made out of discarded fishing nets and line of the kind whales frequently get tangled in by artist Amy Rubin. McDonald devised a series of changing light patterns that are projected onto the sculpture and illuminate the differences in performance by each humpback whale singer. Entering the installation, which has been shown at Mass MOCA in North Adams, Invisible Dog in Brooklyn, Martha's Vineyard, and Cornell University in Ithaca thus far, the viewer feels in the presence of a complex and nuanced artwork, whether or not they choose to read up on what all this means.

Quite simply, it looks and sounds beautiful, and surprising—my criteria for any kind of art. I don't want to see what I expect to see. And I expect a lot because I've seen and heard too much! Aside from being simply an arresting artwork, Lewandowski's careful listening has taught

*Fig. 62:* Siren *as installed at Mass MOCA, 2022 by Annie Lewandowski and Kyle McDonald.*

her another surprising thing: In any one season, the humpbacks do *not* all sing the same song. They each vary their song in quite particular ways, thus being much closer to nightingales and mockingbirds than we originally thought. It may be that the dogma of every male singing the same hip tune, then all changing it together as the season progresses, is a bit of a cliché. Once again we all need to listen more deeply, to hear more than what we expect to hear. Like Maria's deaf students in the Dominican Republic, we must listen outside our ears. We should listen with our mind, with our touch, with our complete senses of wonder.

Like Katy Payne, Annie Lewandowski and Kyle McDonald are *not* ready to publish their discoveries of variation in a single season's humpback song. They need more data and time. Kyle even noticed the kind of group singing Paul Knapp introduced me to the first time I heard humpbacks off the shore of Tortola, still something no scientist has published on:

> This last time that we were at Invisible Dog in Brooklyn, I feel like I started to hear some other things happening in the sound set that I had not recognized before. I think there's a

> harmonic relationship between the different singers. They're not just singing on their own, you know, relative to their own fundamental frequency or something. And I hear different things each time. I'm trying to take this approach, a little bit of like, musical listening, kind of Pauline Oliveros-style deep listening—listening for everything that you can possibly listen for. And then a little bit of listening as a computer might listen, trying to remove your biases, or identify your biases and understand what they are. And then try and step away from them. It's impossible, but it's another way of listening—kind of an *impossible* listening.

We have all been stuck in possible listening, or perhaps biased listening. We've heard the story for so long that all the males in any one ocean are singing the *same* song, we don't listen closely enough to realize it's hardly true. There is so much more variation out there than our old fixed stories will admit. The closer one listens, the more possibilities for new sounds emerge.

In the course of a few decades immersed inside the soundworld of whales, I was approached by the music company Splice, known as the Netflix of sounds, to produce a sample pack of loops, hits, synchronized and unsynchronized, quantized and loose, based on the myriad sounds of whales, from humpback to beluga to narwhal and pilot. I made these several hundred samples together with my son, the hyperpop producer Umru. So now anyone can download a selection of my whale recordings in bite-size format optimized for electronic music production. Sign up and you can try them out here: https://splice.com/sounds/splice-explores/whale-sounds

I am happy that these sounds get to be more widely known and used by more people. If one of my whale recordings turns up on a top ten hit song, that won't bother me. This awareness just might inspire someone to care a little bit more about these great singers in the sea, and that can only be a good thing.

That being said, I doubt I would ever use anything from this sample pack in my own music. I don't like to sample tiny snippets of the music of whales but rather let the songs play on, get used to their rhythm, which

*Fig. 63: Katy Payne and Annie Lewandowski listening to humpbacks in Hawaii (2019).*

is far slower than the usual human pace, and slide my way in. Whatever it takes to get me and everyone else to listen more slowly, more deeply, and more carefully.

Ah but go ahead, search my oeuvre. I know, you'll find plenty of tiny whale, bug, bird, and pond samples. I hope we all dare to contradict ourselves and contain multitudes.

Think of how whale songs have changed humanity! Picture a thriller that begins like this: A submarine is careening back and forth across a rough ocean. Sonar operators are picking up large looming objects and strange sounds that resemble an alien code. Soon they figure it out. "My god, these songs are coming from whales," says one seaman to another. "We can't let people find out about this." And so the navy kept whale sounds from the public for at least fifteen years. Then the secret got out...

Thus, James Bond-like, begins the pitch for the film *The Loneliest Whale*, directed by Josh Zeman, about the mysterious unique 52-Hz sound that is different from any whale species we know. Somewhere between what is expected from fin whales and blue whales, the two largest species, scientists suspected the sound might come from a hybrid whale, an animal never seen before.

I was happy to be part of this project from the beginning, and it took almost ten years to make. Now you can watch it on many streaming services, and find out how the mystery is solved.

Perhaps more important than the solution to this mystery is how the whole story went viral, how moved people all over the world have been by the story of a lone whale crying out for others of his kind who might never have existed. All this technology beautifully connects us while also encouraging a deep sense of emptiness. A paradox?

The sounds of many whale species remain beautiful and mysterious. How can we tell the complex stories of their meaning using only sound, and no explanation? In a way this is the challenge all music faces, though music might not be telling stories at all but conveying more mood, an emotion. Together with Norwegian composer Bjarne Kvinnsland and sound designer Trond Lossius, I was fortunate to work on an ambisonic surround-sound installation for the renovated Whale Hall in the Museum of Natural History in Bergen, Norway.

No one knows exactly how the collection of whale bones got to Bergen, but there they are: skeletons, huge ones, hanging above our heads. A long room 60 meters at least. A nineteenth-century museum, a national landmark for Norway. For seven years the whale skeletons lay enshrouded in great white sheets, protected during the renovation. But be careful if you renovate a landmark, said the authorities. When it is done, it must all look the same. You are allowed to transform it only with sound.

So we got to work. We created a 24-channel dynamic sound installation made out of the songs and calls of sperm whales, pilot whales, killer whales, blue whales, bowhead whales, and an occasional clarinetist trying to make music with humpback whales. The specific aim was a difficult one: to tell precise stories about whale sounds *only* with their sounds, including as little explanatory text as possible. Stories like these:

> **Humpback whales** sing long, beautiful songs that are closer to music than language, sometimes moving human listeners to tears. The male whales change this song, all together, from month to month and year to year. When one whale changes his tune and the new song is accepted, all the other whales quickly follow and change theirs so they all wail the new tune.

When sung deep enough under the ocean's surface, **blue whale** songs can travel thousands of miles across oceans in less than an hour.

**Bowhead whales** sing lilting, simple songs like the bowing of a cello back and forth. They chant as they migrate under the ice. Their sounds suggest ancient time, a slow movement that could outlast human civilization.

**Pilot whales** use a mix of echolocation clicks to find food and whistles and grunts to communicate. Their crisp complicated warbles and thlacks can be easily recorded with an underwater microphone called a hydrophone. Scientists today are trying to analyze hundreds of hours of pilot whale sounds to teach machines to decode what is being said.

**Sperm whales**, the largest of the toothed whales, use clicks for echolocation and also to identify their family groups. They are the only animals we know of that have clan-specific percussive sounds. This precise and cultural way of using sound suggests that these whales have a complex and nuanced social life that humans can barely understand. We knew nothing about this when we hunted them to near extermination for their oil and ambergris.

After three years of transatlantic collaboration, including the design and manufacture of specially created spherical whale bone-tinted speakers, the museum opened to the public in October 2019. A few months later COVID closed it for several years, but I am happy to say it is fully open once again.

On the second night of opening festivities I played a solo concert with our sounds, aware of the irony: How can one encapsulate the grandeur of whales inside a human space? What sense of life could these old bones offer us anyway? Moving between *seljefløyte* (a Norwegian overtone flute), clarinet, and the booming bass clarinet, I interacted with the result of years of our collaborative composition, thinking of the words of Roger Payne, discoverer of the structure of humpback whale song, when he first listened down to it and said, "I heard the size of the whole ocean that night."

The room was dark, the whale sounds moved. I found my way into the mix. When done, everyone in the room smiled. One small way humankind had found a way to touch nature.

The exhibition is permanent—listen to it whenever you happen to end up in Bergen. You can eat whale down at the harbor but you don't have to do that. Just listen to them. And I am honored to say that in part of the exhibition, you can hear me playing live along with the humpbacks too. The new album *Will We Know Why Whales Sing?* (released on the occasion of the publication of this new book) is composed of longer-than-my-usual pieces made up out of the material in the exhibition, combined with slower, more careful solo parts on contralto, bass, and smaller clarinets to represent two decades of immersion in the soundworld of whales. I hope I've learned some humility in all that time—honestly, sometimes I'm still not sure I can get outside my own head and my own ego.

I still don't give a good answer when people ask me how it *feels* to play music live with whales. I never know what to say about my own feelings. When playing this music across species lines, I'm just doing it. I hope it's something larger than myself. I try to enjoy reaching beyond my comfort zone into musics I cannot know. You cannot rehearse for the unknown, like Wayne Shorter says. You can only live in it and surprise yourselves and the whole world. Dare to be unsure, and the world may thank you for it. Bring others along for the journey and they can find their own ways to create at the limits of human understanding too.

Every time I go out to play with whales, interesting things happen. These sounds change people. They change us one way when we merely listen, and they change us again when we dare to join in. Sure, I've cried sometimes when I hear the wails of whales. But I'm too much of a philosopher to dwell simply in my tears. I never know exactly what it all means! I'm never sure how much the whale "gets" me, or if I ever get the whale. I don't want to ask too many questions, since my mind is so trained to swirl around all dialectical possibilities: Does it mean this? Does it mean that? I no longer care what it means, don't you see? I play my way into the mysteries. I just want to keep doing that.

Well, I don't just want to keep doing that. Look at the record before you. I'm still trying to write down the story. You say you're confused!

I'm confused too. Most people get their music today with absolutely *no* information along with it. Physical recordings are dead, liner notes are dead; it's just sound, and image, swirling all around us, competing for our attention.

Playing music live with whales is not like that. It is an experience that cannot easily be captured. Prepare yourself, then just go out there and *do* it.

Late in 2020, heart of the pandemic, I get an email out of nowhere. It's from *National Geographic*, now a subsidiary of Disney. They want to do a foldout graphic whale score and have me play my bass clarinet along exactly with it.

"Do you know I have been trying to get you guys to do a story like for more than ten years?" I wrote to Mesa Schumacher, animal illustrator extraordinaire. "No, we don't know anything about that," she said. "But we need this from you like...yesterday."

Ever since I published the first edition of this book in 2008, at the time called *Thousand Mile Song*, I had been trying to get *National Geographic* to do a story just like this because I knew that in 1979 they included a "sound page" in the back of one of their magazines, a flexible vinyl page that could be put on a turntable to offer up one short fragment of a humpback whale song, one of the most expressive examples of animal music known to humankind. Back then the magazine published so many copies in so many languages that this recording was and is the greatest single pressing of any record ever. They printed ten million copies at once. I'm still amazed by this so I'll tell the story to you one more time!

So in our bean-counting, data-obsessed world one can reasonably say that humpback whales have sung one of the most popular songs ever. Multi-platinum, off the charts!

It's the most popular recorded animal sound of all time. But is it music?

Out of the blue, forty-one years after that famous sound page, *National Geographic* calls me up and asks me to play bass clarinet along with a whale. "Now make sure," says story editor Alberto Lopez a few weeks later, "that you play *exactly* what the whale sings. We don't want any improvisation out of you this time."

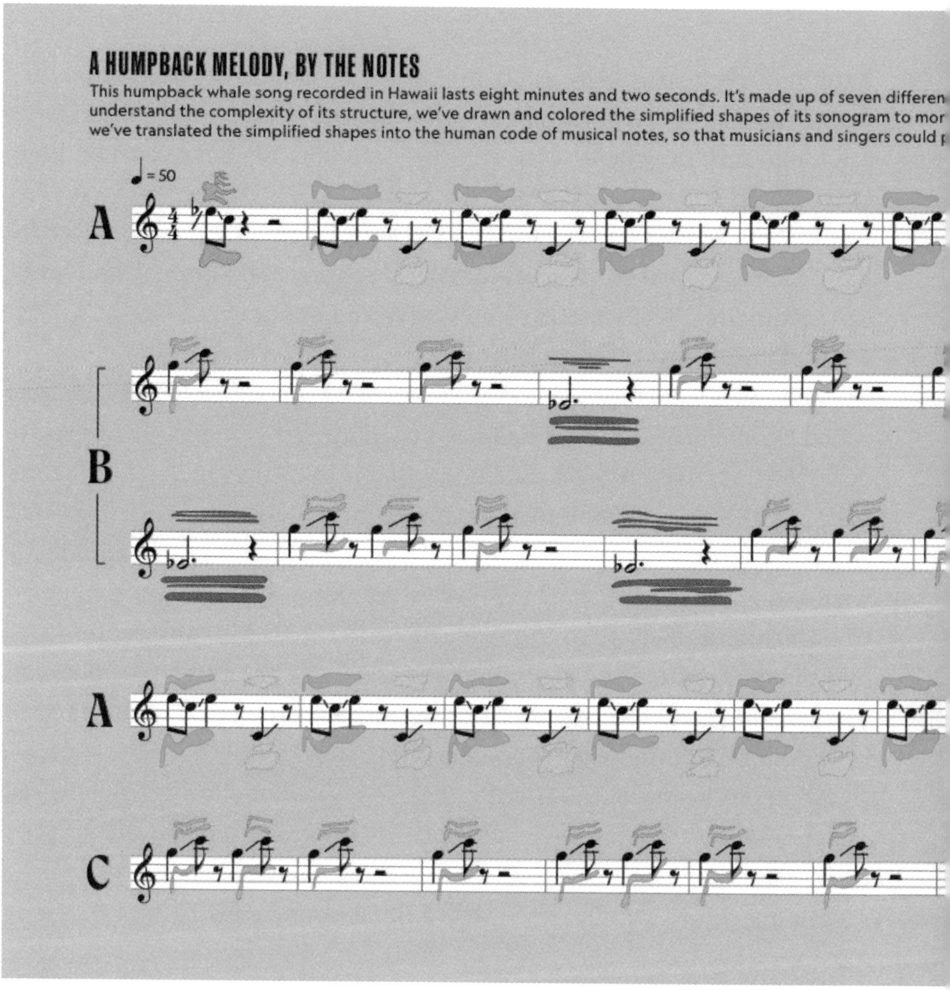

**A HUMPBACK MELODY, BY THE NOTES**

This humpback whale song recorded in Hawaii lasts eight minutes and two seconds. It's made up of seven differen
understand the complexity of its structure, we've drawn and colored the simplified shapes of its sonogram to mor
we've translated the simplified shapes into the human code of musical notes, so that musicians and singers could

♩= 50

A

B

A

C

*Fig. 64 A humpback whale score to be played by all readers of* National Geographic, *May 2021 issue*

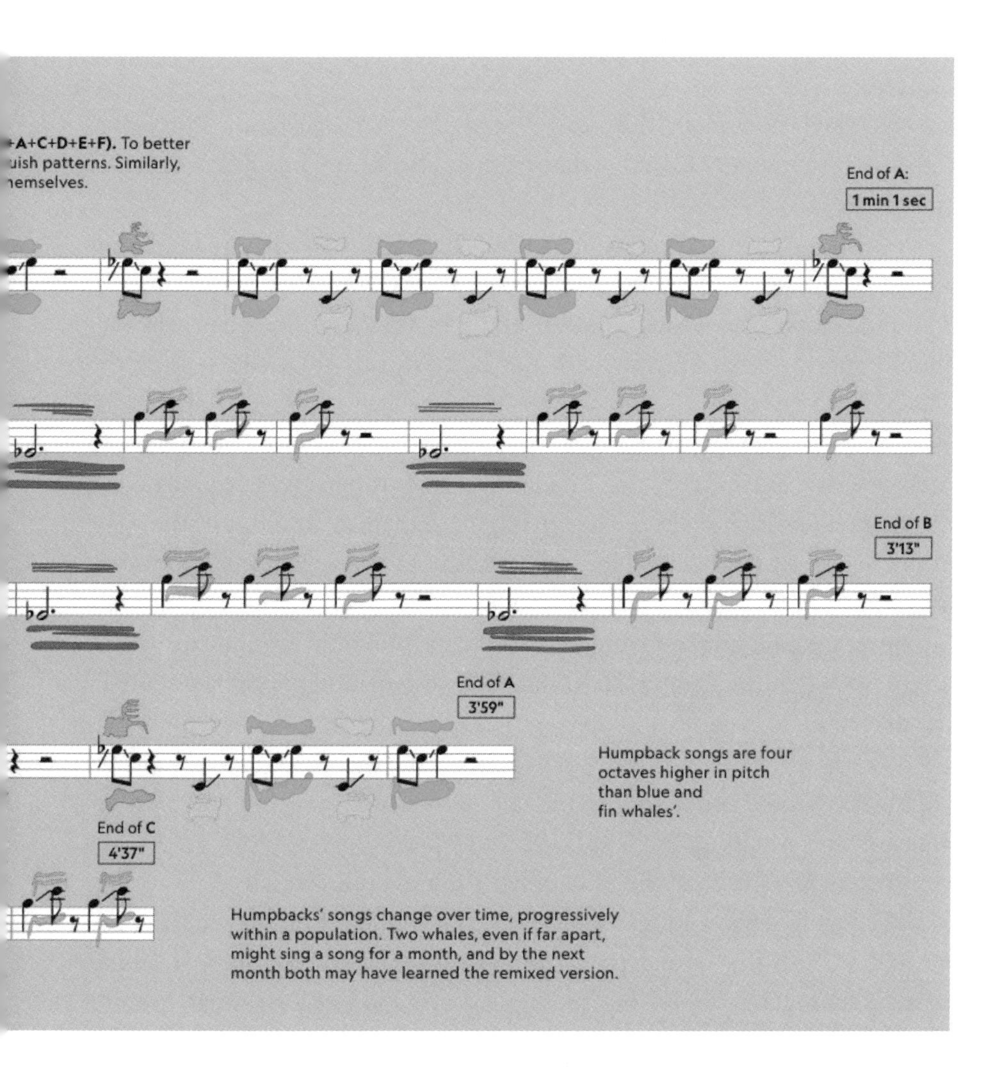
+A+C+D+E+F). To better
uish patterns. Similarly,
hemselves.
End of A:
1 min 1 sec
End of B
3'13"
End of A
3'59"
Humpback songs are four
octaves higher in pitch
than blue and
fin whales'.
End of C
4'37"
Humpbacks' songs change over time, progressively
within a population. Two whales, even if far apart,
might sing a song for a month, and by the next
month both may have learned the remixed version.

Easy, I thought. I've been playing along with whale songs for years. Just like when I play with birds, bugs, and underwater pond lilies, I try *not* to copy what they do but interact with them, create a music with them that neither human, animal, or plant could make alone. Make something together that no species would create apart.

So that's what I did with the whale song. And it wasn't what they wanted.

"No deviation from the score," said the *Nat Geo* team. "Just play right with the whale. Exactly what he does. No funny business." Sink...or swim.

You see, they had a mission. To *convince* their skeptical readers that what the whale was singing was music, not that it could inspire music. So a human playing a low, whale-friendly instrument like a bass clarinet should ape the whale *exactly* with not flourish or new ideas. Just be the whale. Be the whale.

I had to record my part three times before it was close enough to the precise whale music score that I had provided to them. I was used to taking the score like a jazz chart, a place to begin, to inspire, to suggest. But now, at the behest of the biggest nature magazine in the world, I had to follow it exactly, note for note, more closely to the page than any singing male humpback whale himself would do. You can follow the online story here: https://www.youtube.com/watch?v=hNc_Y_Gpitg, and part of the score in the magazine is here:

"Man, that whale sounds more like a bass clarinet than *you* do," said my oldest friend Adam.

"You're absolutely right," I laughed. "Let me tell you why."

The whale sings a note. Then I copy the whale. Then the whale. Then me again. With this cool dynamic interactive notation by Nat Geo artists Mesa Schumacher and Alberto Lopez, based on the original graphic notation devised by Michael Deal and myself. So that, my friend, is why the whale sounds more like a bass clarinet than I do.

He didn't have to follow such instructions, he didn't have to prove anything to the editorial board. He could wail like the whale he is. And here I was, assigned with the task of proving to the readers of *National Geographic* that the humpback whale actually makes music, not noise.

The humpback whale emits "a series of surprisingly beautiful sounds," wrote Roger Payne and Scott McVay in the pioneering cover story "Songs of Humpback Whales" in the journal *Science* in 1971, as I reported in chapter two. They are certainly correct about this. And it is remarkable that no scientific paper since theirs has seen fit to call the humpback song "beautiful" in the many decades since humans discovered the existence of said song. Scientists are taught not to use such subjective or value-laden language these days. It's opening up a can of worms. Or a sea of whales, one way or another.

"You've highlighted the most important line in that paper," said Scott McVay when I spoke to him over the phone when the *National Geographic* story came out. I felt that it was his triumph as well as mine (and the whales') since it was he who got me interested in all this after I first met him at Paul Winter's wedding long ago. I am grateful and honored that he agreed to write a foreword to this new edition of my book.

It was the beauty of the song of the humpback whale that put this great creature on the radar for the environmental cause, setting the tone for the Save the Whales movement that spread across the globe in the 1970s. As I already wrote in chapter two, McVay took a pile of records of *The Song of the Humpback Whale* and played them over national television in Japan to the captains of the whaling industry, who burst into tears exclaiming, "We didn't know, we didn't know."

Beauty can and does make a difference. And I don't mind holding in my own tendency to extrapolate, to improvise, to laugh, to cry.

Yet the whale himself will never hold back. And that is why his song sometimes sounds more like music than mine.

So what's the next step in my collaboration with the music of whales? I still want to get more people out there to join me with their music, fellow musical travelers willing to expand their sensibilities at the edge of our species' aesthetic ways. I want to improve the way whale sounds are visualized, hoping to get together a crack team of the best scientists, artists, and data visualizers to make sonograms greater than ever before. There are ways to push the image envelope that people just haven't thought of, I'm sure.

I'm still hoping to make the best whale music film ever, something that doesn't drench this beautiful natural music in synthesizers and orchestra pads (sorry, Hans Zimmer, we won't need you for this project). Humans have still learned only the tiniest glimmer of the possibilities inherent in the magic of these great whales. "Get inside the whale," implored George Orwell, using whales as a metaphor for the whole predicament of our society. "Or admit," he continued, "that we're all already inside the whale," and there probably is no way out.

Sure, our human/nature continuum has a lot to answer for, problems piling up on top of each other that even a few decades back we could barely dream of.

Why care about whale songs in such a time of strife?

As the climate changes, whales to must find a way to thrive. In 2019 even the International Monetary Fund published an article describing great whales as "Nature's Solution to Climate Change" and as "an international public good," noting that the largest animals thrive in tandem with some of the smallest: "In recent years, scientists have discovered that whales have a multiplier effect of increasing phytoplankton production wherever they go. How? It turns out that whales' waste products contain exactly the substances—notably iron and nitrogen—phytoplankton need to grow."

In a world where big decisions seem to be made mostly on economic grounds, the IMF notes that phytoplankton productivity, which is enhanced by baleen whales, captures 37 billion tons of $CO_2$ per year! Continuing their line of economic thinking, they argue as follows:

> We estimate the value of an average great whale by determining today's value of the carbon sequestered by a whale over its lifetime, using scientific estimates of the amount whales contribute to carbon sequestration, the market price of carbon dioxide, and the financial technique of discounting. To this, we also add today's value of the whale's other economic contributions, such as fishery enhancement and ecotourism, over its lifetime. Our conservative estimates put the value of the average great whale, based on its various activities, at more than $2 million, and easily over $1 trillion for the current stock of great whales.

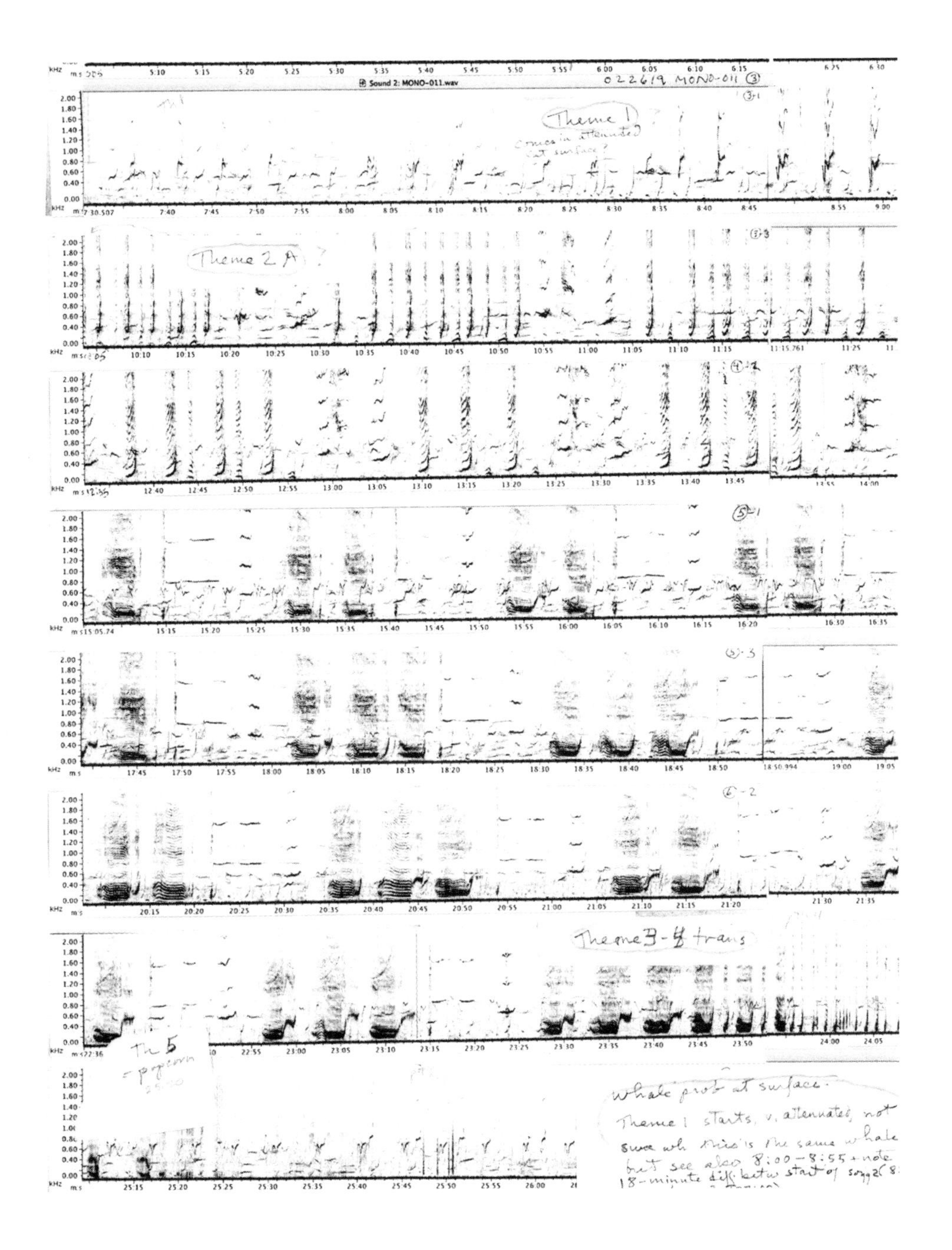


*Fig. 65 Katy Payne's hand-annotated humpback sonogram, 2019 season.*

Save whales for the profit, and save the planet while you're at it!? This is a far cry from how we began this chapter, arguing that we should value animals based on their unique capabilities, and that is certainly what got me excited about whale music in the first place. It is actually the *sound* itself that moved me to tears of wonder, and my own hubris that I had any sort of right to join in. Don't give in to the money...

I'm still hoping to get back to that beach in Tortola to ask Paul Knapp all these questions. I will apologize to him that it's taken so long! I know he's smiled when he has seen how far his ultimate Valentine's Day whale song has traveled. Let it inspire a whole new generation of singing whales and humans and activist musicians from all cultures finding their own way to make a difference in this important time when our own species will be judged against the planet, and who might ask: Are we better off with humans, or without?

At least the better humans know to listen to the whales, and dare to join in.

Thanks for following along with me on this journey.

# ACKNOWLEDGMENTS

The search for whale sounds is a difficult quest that takes would-be listeners out to sea for weeks or months on end, with much effort and patience. The leading figures in the field who appear in this book were all open, helpful, and eager to consider my nontraditional approach. Thanks especially to Paul Knapp, Scott McVay, Chris Clark, Vincent Janik, Hal Whitehead, Katy Payne, Michel André, Mark Johnson, Peter Tyack, Margaret Lovatt, Sal Cerchio, and particularly Jim Darling, who got official permission for me to join him out in Hawaii aboard his research vessel, the *Never Satisfied.*

Thanks to musicians Pete Seeger, Paul Winter, Michael Fahres, George Crumb, Alex South, Annie Lewandowski, and Abigail Sanders for taking the time to speak with me about their work and how whales fit into it.

My first three main expeditions in the field to make music with whales were made possible through the help of many encouraging individuals. In British Columbia, Jim Nollman and James O'Donnell made the journey a success, even if two of our boats broke down. Ken Ramirez let me into the Shedd Aquarium in Chicago for three days to make music with their belugas in preparation for the journey to Karelia in Russia, which would never have been possible without the persistence of Rauno Lauhakangas and the company of Gari Saarimaki, Anna Koivisto, Aleksandr Velikoselsky, and the Lechki brothers of Kolezhma, who even built a brand-new sauna on Myagostrov Island in honor of our arrival. In Moscow, Aleksandr Agafonov and Roman Belikov were full of enthusiasm and ideas. Olga Filatova shared with me her vast electronic collection of whale science papers, direct from Kamchatka.

The trip to Hawaii would never have been successful without the support of Dan Sythe and Kent Noonan of the Whalesong project. Meagan Jones and Flip Nicklin of the Whale Trust organization also graciously brought me into their world of research and photography. Willy "The Whale" Bennett and Heather Harding brought science and the mystical together. Sheryl O'Day graciously offered her catamaran for a day. Musicians on Maui including Don Lax, Ashana Morrow, and Lauren Pomerantz shared their thoughts openly on music and whales. Casey Robinson kept my spirits up.

Friends and colleagues who read portions of the book and came up with leads and links from music and science into the world of cetacea include Sylvia van der Woude, Helen Arusoo, Jim Cummings, Doug Quin, Richard Nunns, Adam Clayman, Scott Diel, Michael Pestel, Jay Griffiths, David Abram, Dario Martinelli, Daniel Rothenberg, John Brien, Erich Hoyt, Tecumseh Fitch, Sam Bower, Nils Økland, Jenny Toomey, Martin Nweeia, Doug Chadwick, Michelle Makarski, Petri Kuljuntausta, Robert Jürjendal, Ann Warde, Morton Subotnick, Eliza Michalopoulou, Jennifer Jackson, Mark Pilkington, Samone Yust, Ken Jordan, Marilyn Crispell, D. Graham Burnett, Leesa Sklover, Bill Rossiter, Christina Coffman, John Wieczorek, Wendy Lindbergh, Jennifer Sahn, Robin Rimbaud, Danielle Cholewiak, Renata Sousa-Lima, Jan Bang, Lisa Walker, Jim Motavalli, Andy Revkin, Bridget Nicholls, and Lawrence Weschler.

Over the fifteen years since the publication of the first edition of this book, I have met many fellow whale enthusiasts who have helped me expand this project toward a future that is still unknown. These include Nicole and Alexander Gratovsky, Oriana Kalama, Josh Zeman, Tilman Farley and Susana Yalala Gonzalez, Olivia Wyatt, Daniel Opitz, Rick Bass, W.S. Merwin, Tristan Visser, Marie-Kell de Cannart, Olivier Adam, Michael Stocker, Mark Fischer, Carl Safina, Aline Penitot, Bernard Abeille, Radha Ala, Garth Stevenson, Alexandra Morton, Heike Vester, Heather Woodson-Gammon, Bjarne Kvinnsland, Trond Lossius, Gaelin Rosenwaks, Arkady Shilkloper, Camille Hanson, Roger Payne, Peter Gabriel, Laurie Anderson, Sue Savage-Rumbaugh, Kyle McDonald, Alex South, Emily Doolittle, Annie Lewandowsi, and Abigail Sanders.

Thanks to the New Jersey Institute of Technology for granting me sabbatical leave to work on a project that would not have been possible without such unobstructed time. This is an institution forward-looking enough to accept whale music as a proper subject for study by a member of its humanities department. I am grateful to the Safina Foundation for three years of generous support in recent years.

Wandee Pryor did exhaustive background research and preliminary editing of the first edition. Thanks to Serge Masse and Martin Pedanik for the conception and design of the original diagrams, and to Martin for his exemplary design of this new book after years of putting up with my complaints. Thanks to Wade and Robyn Hughes for being so generous to allow me to use several photos from their beautiful book *Looking for Whales* (Sydney: Halstead Press, 2019). For the new edition years of work with visualizer and designer Michael Deal have been essential, and the editorial advice of Tyran Grillo, Evan Eisenberg, Colette McCormick, and John P. O'Grady of Terra Nova Press and Marc Lowenthal from our distributor MIT Press.

Thanks to my family for all their understanding and support in the years I have spent obsessed with the far-flung topics. Over time they do start to seem less obscure. The world listens to the world, and hopefully dreams ever more of saving it. I hope the stories brought back have been worth the faraway journeys to the mysterious sounds of the deep.

# WHALE MUSIC: THE SOUNDS

The original 2008 edition of this book included a CD bound beautifully into book in a plastic sheath. How times have changed. Now in 2022 all music streams freely on all the currently popular platforms that pay their creators almost nil. These platforms strip music of all description, all context, all information. So I present the liner notes here for the various whale-related albums I have released. These are:

*Whale Music* (2008)
*Whale Music Remixed* (2009)
*New Songs of the Humpback Whale* (2015)
*Who Knows Why Whales Sing?* (2022)

It's a lot of detail, but I hope you find it useful. The music should be easy enough for you to find online, but music online is released "in the context of no context." Lacking physical form, how much do you think it's worth?

Remember these beasts have been singing such music for millions of years. Surely that alone should be worth something to we mere humans!

Here is a playlist with all of the music described below.

## *Whale Music* (2008)

*Whale Music* is the first musical record of my journey into the sound world of whales, released directly to correspond with the tales in this book. Some pieces were produced in the studio, presenting music made out of their structure and tones, showing how we humans can learn from this deep sea alien music. Others are unedited, live encounters between clarinetist and whale. These very different approaches offer two distinct ways for a human musician to delve into the possible musical inspiration offered by the cetacean world.

**Track 1. Valentine's Day, 1992**

David Rothenberg, clarinet
Robert Jürjendal, guitar
Humpback whale recorded by Paul Knapp off Tortola

Here is the best whale song recording Paul Knapp ever made, used as the basis for a trio with two humans joining in. The whale recording is slowed down and transposed a fifth down, so that the structure of his song is clear. Most human music based on whale songs concentrates on the timbre of the wail, not the form, with its pattern and rhyme, which is the most amazing thing about the music of the humpback whale.

**Track 2. Never Satisfied**

David Rothenberg, clarinet
One close and one far humpback whale
*Recorded live in February 2007, off the coast of Maui*

This is my favorite live unedited duet of me playing along with humpback whales. It seems as if at least one whale is reacting to me, changing his pitches and rhythms in response to mine. This may be wishful thinking, but on the other hand, if the whales are able to rapidly change their songs over even just a few weeks in response to brand-new songs, one would expect them to be able to immediately respond to a whalesque clarinet sound as well.

**Track 3. The Far Field**

Nils Økland, violin
David Rothenberg, clarinet, electronics
Samples of ruffed grouse, catbird, woodpecker, humpback whale
Published by Mysterious Mountain Music (BMI) and Nils Økland (TONO)

Nils Økland's violin playing is heavily influenced by the Hardanger fiddle, a Norwegian folk violin with sympathetic strings like a sitar. I've lowered the pitch on it here so it becomes more like an imaginary Hardanger cello. Similarly, the slowed-down catbird sounds like a humpback whale, and the grouse and woodpecker sound like sperm whale clicks. The real whale rises faintly in the distance.

**Track 4. Whiteness of the Beast**

John Wieczorek, percussion
David Rothenberg, bass clarinet, synthesizer
Beluga whale sounds recorded by Roman Belikov, Solovetski Island, White Sea

A duet between percussion and bass clarinet is enhanced with the ambient, deeply transposed repertoire of the white beluga whales of Russia's White Sea.

**Track 5. Duo Orcananda**

David Rothenberg, clarinet
Northern resident orcas of the A1 pod, Vancouver Island
*Recorded live in September 2006, Johnstone Strait*

Live interaction between clarinet and killer whales, several miles away. The whole thing is transposed down an octave to be more listenable.

**Track 6. The Killer**

Robert Jürjendal, guitar
David Rothenberg, clarinet
Calls of mammal-eating transient orcas, Vancouver Island

These are the sounds of those orcas whose culture tells them to eat other mammals, not fish. Here is the melancholy lament of those destined to feed closer to their own kind: on seals, otters, deer, moose, even other whales. That's what their families teach them to do.

**Track 7. Moby Click**

David Rothenberg, bass clarinet, synthesizer
Resonated clicks of sperm whale codas recorded by Hal Whitehead

The clicking overlapping coda rhythms of female sperm whales can be turned into a beat that sounds like it's played on a Brazilian berimbau, which is a one-string buzzy guitar played with a coin.

**Track 8. And She Married a Whale**

Michelle Makarski, violin
David Rothenberg, clarinet, electronic treatments

No real whales here at all, just humans wrapped up in the cetacean aesthetic.

**Track 9. Myagostrov, in the Deep**

David Rothenberg, clarinet
Belugas of the White Sea
*Recorded live in July 2006, Myagostrov, Karelia*

Reaching out live to the belugas of the White Sea, we hear the clacking rhythms of the rough surf on the rocks and the broadband beluga noises near the hydrophone, blaring in and around the clarinet, a window into a foreign conversation whose meaning we can only guess at.

**Track 10. Beluga No Believe in Tears**

David Rothenberg, Turkish G clarinet, keyboards

Beluga whale sounds recorded by Roman Belikov, Solovetski Island, White Sea

The rhythms and stereo horn moans are all derived from original beluga recordings.

**Track 11. Koholaa!**

David Rothenberg, bass clarinet

One male humpback whale, singing close to the shore of Kihei, Maui

A carefully transformed whale song becomes the basis for this composition. Compare the structure and shape of this Pacific whale song with the Atlantic whale song in track 1.

**Track 12. The World's Last Whale**

David Rothenberg, vocals, Turkish G clarinet, synthesizer

Words and music by Pete Seeger © 1970 Storm King Music Inc. (BMI)

Rhythm and background derived from the sped-up rhythmic calls of minke whales, blue whales, and fin whales, recorded by the U.S. Navy (declassified under the Dual Uses Initiative of 1991); Belikov's beluga screams also heard.

Here is the first published recording of Pete Seeger's classic song about how recording the sounds of whales can inspire us to save them. I couldn't talk Pete into singing it for us, so I had to do it myself.

Produced by David Rothenberg.
Recorded on location in Hawaii, Karelia, and British Columbia.
Studio recordings at B Street Studios, Cold Spring, New York, and Pune Tee, Tallinn.

Special thanks to Jim Nollman, Rauno Lauhakangas, Gari Saarimaki, Jim Darling, Dan Sythe, Kent Noonan, Aleksandr Agafonov, Roman Belikov, Paul Knapp, Hal Whitehead, Edie Meidav, Chris Clark, and Pete Seeger.

Michelle Makarski appears courtesy of ECM Records.

Nils Økland appears courtesy of Rune Grammofon.

David Rothenberg
Whale Music Remixed
by
Scanner
3Corners of the World
Gari Saarimäki
Mira Calix
Ben Neill
DJ Spooky
Warren Burt
Markus Reuter
Lukas Ligeti
Robert Rich
White Sea Shamans
Francisco Lopez
Strings of Consciousness
Stephen Chopek

## *Whale Music Remixed* (2009)

Why remix a whale sound? Isn't it fine just as it is? In today's fluid world of music, no noise is innocent, no sound is safe from being turned into any other. These undersea tones of mystery, whose meaning is hardly understood by humans, suggest all kinds of creative possibility the minute you begin to listen deeply and to inhabit them.

If you think you know what a whale sounds like think again. Their voices offer more than the famous long, deep moans of the humpback strung together in long, structured songs. There are the Morse code-like clicks of sperm whales, and the squeals and echolocation creaks of belugas. The cries of orcas, and subsonic booms of the largest fin and blue whales, deep sounds that can cross whole oceans in less than an hour.

Last year I released a recording called *Whale Music*, where I explored the sonic possibilities of these great beasts, live in the ocean and in the studio.

This year I sent copies of this record to the finest electronic musicians I know, with hopes that they might do great things with such unusual material. The contributors come from the far corners of the Earth, making music of many kinds, all infused with the depth and wonder of the songs of whales. Markus is in Germany, Warren in Australia, Cycle and the Shamans in Russia, Strings of Consciousness in France, Francisco in the Netherlands, 3Corners in Estonia, Mira and Scanner in the UK, and the rest of us in the USA.

Two pure whale recordings are included for your own remixing pleasure, belugas from Russia's white sea, and one lone male humpback off the shore of Maui smack in the middle of mating season.

A portion of the proceeds from this recording was used to benefit the Whalesong Project, which broadcasts humpback whale sounds live from Hawaii over the internet, www.whalesong.net

—David Rothenberg, April 2009

**1. Killa Koholaaa (5:04)**
Paul D. Miller aka DJ Spooky
Subliminal Kid Publishing (BMI) / Mysterious Mountain Music (BMI)

**2. mr moby declicked (4:27)**
Markus Reuter
Markus Reuter (GEMA) / Mysterious Mountain Music (BMI)

**3. Valentine's Whale Remix (3:56)**
Robert Jürjendal and David Rothenberg (3Corners of the World)
Mysterious Mountain Music (BMI)

**4. The Far Field (WNYC Remix) (4:14)**
Scanner and David Rothenberg
Scannerdot Publishing (Bug Music PRS) / Mysterious Mountain Music (BMI)

**5. SodaCan Whale (4:17)**
Stephen Chopek
Stephen Chopek (BMI)

**6. Belugas of the White Sea (1:55)**
the complete unaltered vocabulary of the beluga whales of Russia's White Sea recorded by Roman Belikov

**7. Call Dem Whales (1:55)**
White Sea Shamans
Mysterious Mountain Music (BMI)

**8. Beluga No Tears (5:07)**
Gari Saarimaki and David Rothenberg
Mysterious Mountain Music (BMI)

**9. &shemarriedawhale (5:20)**
Mira Calix
mutesongs (PRS)

**10. cAChaLotic trialogue (3:04)**
Lukas Ligeti
Lukas Ligeti (AKM) / Image of Two Ears (ASCAP) / Mysterious Mountain Music (BMI)
features the click sounds of Physty the sperm whale recorded by Bill Rossiter

**11. Last Whale Remix (2:40)**
Cycle Hiccups
Mysterious Mountain Music (BMI)

**12. Beluga Hummingbird (5:03)**
Warren Burt
Warren Burt (APRA)
features the voice of John Lilly and the treated song of a hummingbird

**13. Whale Lake (6:42)**
Strings of Consciousness
Perceval Ballone and Philippe Petit (SABAM)
Perceval Bellone (saxophone/EWI), Philippe Petit (laptop/turntables)

**14. F**king with Animals (4:02)**
David Rothenberg and Mark Johnson
Mysterious Mountain Music (BMI)
featuring the voice of Mark Johnson

**15. untitled #219 (4:12)**
Francisco López
created at mobile messor (stockholm)
Mysterious Mountain Music (BMI)

**16. Myagostrov (Hudsonia Remix) (4:49)**
Ben Neill
Blue Math Music (ASCAP)
Ben Neill: mutantrumpet/electronics

**17. One Lone Maui Humpback Whale (4:23)**
recorded by David Rothenberg

**18. Spectral Canyon (4:58)**
Robert Rich
Amoeba Music (BMI)

musicians on the original *Whale Music* CD (TN-0804) who are remixed here:

David Rothenberg, clarinet, bass clarinet, soprano sax, whale soundscapes
Michelle Makarski, violin
Robert Jürjendal, guitar

NEW SONGS OF THE
HUMPBACK WHALE

## *New Songs of the Humpback Whale* (2015)

Compilation of original humpback whale recordings
by the best listeners and scientists out there.

by David Rothenberg and Michael Deal
Important Records, Newbury, MA

What does it take to record the best whale songs? Technology isn't all that you need. One must have time, a lot of it, to go out on the water, drop your hydrophone deep down, get ready to listen, and to wait.

Upon hearing the great song for the first time, Roger Payne said he heard the size of the ocean, "as if I had walked into a dark cave to hear wave after wave of echoes cascading back from the darkness beyond.... That's what whales do, give the ocean its voice." Most work on the meaning of these tones is far more prosaic, involving pages of calculations, summary charts that have a hard time containing the original beauty....

One scientist who is committed to working on humpbacks is Olivier Adam, at the Sorbonne in Paris. He is also a researcher who works in the spirit of Payne and McVay, in that he appreciates the beauty of the song and has worked for years in the muddy waters off Madagascar to get the best possible recordings. When not working on whales he is also involved in research on the indigenous music of this island nation, and he is not afraid to tackle some of the toughest problems in humpback song research. Such as: exactly *how* do humpbacks make such tremendous songs? It turns out no one really knows. No humpback has survived in captivity long enough for anyone to examine the process closely. But Olivier and his team have just published the first attempt to model the song production process that *might* explain how the whales do it. No air leaves the whale while he sings.

Our selection of songs included come from many places. Glen Edney, formerly a whale watching tour operator in Tonga, contributed one beautiful hour long song. Sal Cerchio, a researcher at the Bronx Zoo's Wildlife Conservation Society, offered the best of his several decades of African coastal recordings. I offer my own recording made unusually in Hawaii in very shallow, muddy water, just after a storm, when one lone male was singing close to the shore in Kihei. These conditions explain why there is very little background noise on this recording, and what little there was I took out. I include a second recording from roughly the same spot, four years later. Can you hear any changes in the song?

Why take out the noise on underwater field recordings you might ask? Underwater can be a noisy place, especially from the perspective of a hydrophone, which can pick up the ubiquitous crackling noise of snapping shrimp, the thrum of boat engines up to ten kilometers away, and the rubbing of cables and anchors against the boat itself. Sometimes there are so many whales singing it can be hard to isolate the pattern and structure of just one. In fact most underwater whale recordings feature many whales, and sound like overlapping insect choruses more than clear solos. But noise reduction software is pretty advanced today, and can almost automatically remove continuous frequencies that get in the way of what we want to hear. I favor those that work in real time, rather than the highest quality methods that separately process the sound.

But when you take *all* the noise out of a recording it suddenly loses its life. It sounds like a technical document, a test of machine ingenuity. Of course every recording is something artificial, a confluence between a phenomenon in the world and the latest human machinery that tries to capture it. It's all illusion, and some illusions can be better than others. I do think some recordings can be more honest than others, and these are not always the ones that have the least processing on the sound. One works with what the hydrophone has brought in, and massages the sound to reveal the beauty and detail that lies within.

And of course if these songs are literally sung for *hours*, how do you decide how much to release? Sometimes I think of really offering up many hours of a single performance, just to encourage the human audience to get into whale time, to slow down, to try to take the whole song in at an extended scale beyond the usual senses of our own species. In concerts when I play along with whale song I usually do it for only five to ten minutes, but I *want* to do it for at least thirty minutes, maybe a whole hour, to really change the way the audiences senses what more-than-human music can be. But I rarely dare to.

On our recording we strive for a balance. Five long whale songs, each nearly ten minutes in length. Some of these sounds strain the limits of human aesthetic sense. Booming lows too deep to relax with, leaping to sudden high screams. Vast contrasts in emotions. The sixth song is another of Olivier Adam's Madagascar recordings, but this one played at double speed, while simultaneously transposed down an octave and a half, hoping this double transformation will get the song closer into the human range of hearing and rhythmic comprehension.

A few years ago humpback brains were found to contain a kind of cell called a spindle neuron that previously was only known to appear in the brains of higher-order primates, those animals thought to be able to experience complex emotions. Once more whales join a small club of which we and chimpanzees are both members. If they can do it, it is most likely in these deep and complex songs that the whales let loose the widest range their feelings can contain. Sit back and listen, take in the strange and the jarring. Try to delve deeper into the whales' own songs, to present them for others to hear with the greatest respect, space, and sound quality as possible. They *do* sound different than the first songs recorded in the sixties. To every generation they will sound different again, as our ears, and the whale's songs, are constantly on the move.

—David Rothenberg, September 2014

| | |
|---|---|
| **1. Madagascar Whale One**<br>recorded by Salvatore Cerchio 1990 | 10:09 |
| **2. Madagascar Whale Two**<br>recorded by Olivier Adam, 2007 | 10:29 |
| **3. Tonga Whales**<br>recorded by Glenn Edney, 2008 | 9:45 |
| **4. Maui Whale One**<br>recorded by David Rothenberg, 2007 | 11:41 |
| **5. Maui Whale Two**<br>recorded by David Rothenberg, 2010 | 8:01 |
| **6. Madagascar Whale Lower, Faster**<br>recorded by Olivier Adam in 2007 | 5:12 |
| TOTAL TIME | 55:13 |

produced by David Rothenberg
artwork and visualization by Michael Deal

Important Records
Newbury, MA

all titles published by Mysterious Mountain Music BMI
no animals were harmed in the production of this recording

Why ?

## *Who Knows Why Whales Sing?* (2022)

released in November 2022 on the occasion of the publication of this book.

### 1. Clarinet and Humpback Day 5 (5:52)

Humpback whales of the shores of Maui, March 2019
David Rothenberg, clarinet

Recorded by Bjarne Kvinnsland, live and unedited

This is the last time I went out to play live with humpback whales. I miss it. We'll be back soon.

### 2. Of Killers and Clarinets (9:48)

Killer whales (Orcas), recorded by Heike Vester, Lofoten Islands, Norway, 20??
David Rothenberg, contralto clarinet, electronics

Where they go high, I go low. Lower and lower with each passing year. Still not low enough.

### 3. My Pilot, Whale (3:14)

Pilot whales, recored by Heike Vester, Lofoten Islands, Norway, 2010
David Rothenberg, *fujarka* [custom Czech overtone flute by Marek Gonda], electronics

The whales themselves sound ever more electronic at these frequencies, pushing me into the beyond.

### 4. Will We Know Why Whales Sing? (10:12)

Humpback whale, recorded by David Rothenberg, Maui 2010. Electronically transformed. David Rothenberg, clarinet, contralto clarinet.

You got to spend time with these songs. Inhabit them. Usually I shorten everything to appease human listeners. Not this time.

**5. The Great Cachalot (9:13)**

Physty the sperm whale, recorded in 1981 By Bill Rossiter. Remixed.
David Rothenberg, contralto clarinet, electronics

The sounds of sperm whales are used for echolocation, and also social communication, perhaps closer to language than to music. Here I push them back to the realm of pure sound.

**6. Thousand Mile Songz (6:20)**

Fin whale whoops, courtesy NOAA
David Rothenberg, contralto clarinet, electronics

These deep low sounds, usually heard several minutes apart in an astonishingly regular beat, are here sped up so we humans can feel their pulse too. Listen on a system with lotsa bass.

**7. Belugas Across Time (5:35)**

Beluga whales, recorded by Roman Belikov in Kamchatka, 2004, looped and remixed. David Rothenberg, Norwegian bass *seljefløyte* by Jean-Pierre Yvert, electronics

The echolocating belugas looped form the rhythm here. Above and below we try to fill their world in.

**8. Clarinet and Humpback Day 2 (6:24)**

Humpback whales of the shores of Maui, March 2019
David Rothenberg, clarinet

Recorded by Bjarne Kvinnsland, live and unedited

Does the whale hear what I play to him? Yes. Does he care?
Still no one knows why whales sing.

Perhaps we ought to keep it that way.

TOTAL TIME: 57:00

Produced by David Rothenberg
Released November 2022
Surround sound mix expected in Spring 2023



TN2227
www.terranovamusic.net

# FOR FURTHER READING

Whales live far away from where people usually do, so we know little about them. But that hasn't stopped us from writing many books about killing them, communicating with them, and just trying to find them across the vast oceans.

The adventurous but bloody history of whaling has inspired numerous exciting accounts, going back to the early nineteenth century. William Scoresby, *The Northern Whale-Fishery* (London: Religious Tract Society, 1820), is the first famous one, and Charles Nordhoff, *Whaling and Fishing* (New York: Dodd Mead, 1895 [originally published 1856]) was far more popular in its day than *Moby Dick.* Charles Scammon's *The Marine Mammals of the Northwestern Coast of North America* (New York: Dover 1968 [originally published 1874]) is the most scientific of these early books, and it is instructive to look back at Adriaen Coenen, *The Whale Book* (London: Reaktion, 2003 [originally published 1585]), to see how close the undersea world was to fantasy way back then. These historical works have only scattered references to the sounds of whales, if they mention them at all.

Among modern encyclopedic works on the relationship between humanity and cetaceans, Richard Ellis, *Men and Whales* (New York: Knopf, 1991), is without peer. A much more compact, cultural vision is found in Joe Roman, *Whale* (London: Reaktion Books, 2006). The best book on the way humans began to consider whales in an entirely new way after the discovery of humpback song is the unique anthology *Mind in the Waters,* edited by Joan McIntyre (New York: Scribner's, 1974), with contributions by Paul Spong, John Lilly, Scott McVay, Gregory Bateson, Victor Sheffer, Peter Warshall, Farley Mowat, and many others.

Other excellent books on whales include Douglas Chadwick, *The Grandest of Lives: Eye to Eye with Whales* (San Francisco: Sierra Club Books, 2006); Kenneth Brower, *Freeing Keiko: The Journey of a Killer Whale from* Free Willy *to the Wild* (New York: Gotham, 2005); Peter Heller, *The Whale Warriors* (New York: Free Press, 2007); and James Darling, Flip Nicklin et al., *Whales, Dolphins and Porpoises* (Washington, DC: National Geographic, 1995), which is one of my favorite whale coffee table books. Another is

Bill Hess, *Gift of the Whale: The Iñupiat Bowhead Hunt* (Seattle: Sasquatch, 1999), one of few picture books to take the whalers' point of view.

Since the publication of the first edition of this book in 2008 so many extraordinary books on whales have come out, notably Philip Hoare, *The Whale* (New York: Ecco, 2011) and *Albert and the Whale* (London: Fourth Estate, 2021), D. Graham Burnett, *The Sounding of the Whale* (University of Chicago, 2012), Rebecca Giggs, *Fathoms: The World in the Whale* (New York: Simon and Schuster, 2020), Joshua Horwitz, *The War of the Whales* (New York: Simon and Schuster, 2015), Margret Grebowicz, *Whale Song: Object Lessons*, (New York: Bloomsbury, 2017), Alexis Pauline Gumbs, *Undrowned: Black Feminist Lessons from Marine Mammals* (Chicago: AK Press, 2020), Nick Pyenson, *Spying on Whales* (New York: Penguin, 2019), Michael J. Moore, *We Are All Whalers* (University of Chicago Press, 2021), Ryan Tucker-Jones, *Red Leviathan: A Secret History of Soviet Whaling* (University of Chicago Press, 2022), and Tom Mustill, *How to Speak Whale* (New York: Grand Central, 2022).

There are now a few decades worth of large scientific anthologies on whales. Kenneth Norris, ed., *Whales, Dolphins, and Porpoises* (Berkeley: University of California Press, 1966), is quite comprehensive, though understandably out of date. Seventeen years later came Roger Payne's *Communication and Behavior in Whales* (Boulder, CO: Westview Press, 1983), which contains key papers from the first decade of humpback and orca sound research. *Dolphin Societies* (Berkeley: University of California Press, 1991), edited by Karen Pryor and Kenneth Norris, includes much valuable research, including translations of important Russian papers. The recent *Cetacean Societies* (University of Chicago Press, 2000), edited by Janet Mann, Richard Connor, Peter Tyack, and Hal Whitehead, emphasizes the group aspect of whales' lives. Denise Herzing, *Dolphin Communication and Cognition* highlights more recent work on cetaceans in the wild (Cambridge: MIT Press, 2015), and Janet Mann's edited anthology *Deep Thinkers: Inside the Minds of Whales, Dolphins, and Porpoises* (University of Chicago Press, 2017), is a beautifully illustrated half-academic half-coffee-table whale book.

Several scientists have written fine memoirs on their experiences in the field, including Pierre Béland, *Beluga: A Farewell to Whales* (New York: Lyons & Burford, 1996), Erich Hoyt, *Orca: A Whale Called Killer* (Tonawanda,

NY: Firefly Books, latest edition 2019), Jim Darling, *Hawaii's Humpbacks* (Vancouver: Granville Island, 2009), Flip Nicklin, *Among Giants: A Life with* Whales, Chicago: University of Chicago Press, 2011), Diana Reiss, *The Dolphin in the Mirror* (New York: Mariner, 2012), and Roger Payne, whose incomparable *Among Whales* (New York: Scribner's, 1995) is a veritable Leviathan of a book by the best public advocate the whales could ever ask for. Hal Whitehead's *Sperm Whales: Social Evolution in the Ocean* (University of Chicago Press, 2003) is in a class by itself—a serious science book about a single whale species that is also tremendously engaging for the non-specialist reader. The sequel co-authored with Luke Rendell is the definitive statement on *The Cultural Lives of Whales and Dolphins* (University of Chicago Press, 2015).

On the subject of human communication with cetaceans, Toni Frohoff and Brenda Peterson, *Between Species: Celebrating the Dolphin-Human Bond* (San Francisco: Sierra Club Books, 2003), shows the wide range of what is possible between us and them. From the traditional animal-training perspective, Karen Pryor, *Lads Before the Wind: Diary of a Dolphin Trainer* (North Bend, WA: Sunshine Books, 1989 [originally published 1975]), shows how much these animals can be taught, and Ken Ramirez, ed., *Animal Training: Successful Animal Management Through Positive Reinforcement* (Chicago: Shedd Aquarium, 1999), is the encyclopedia on how it should be done.

When it comes to making music with whales, Jim Nollman has been at it longer than anyone. *The Charged Border: Where Whales and Humans Meet* (New York: Henry Holt, 1999) remains his best book, though one should not overlook *The Beluga Café* (San Francisco: Sierra Club Books, 2002). On the idea of music among animals in general, see François-Bernard Mâche, *Music, Myth, and Nature; or The Dolphins of Arion* (Chur, Switzerland: Harwood Academic, 1983), or the other foundational tract of "zoomusicology," Dario Martinelli, *How Musical Is a Whale?* (Helsinki: Acta Semiotica Fennica XIII, 2002). Nils Wallin, Björn Merker, and Stephen Brown, eds., *The Origins of Music* (Cambridge: MIT Press, 2000), gives a range of reasons for how music evolves through nature into humanity, and since then we got Tobias Fischer and Lara Cory, *Animal Music* (London: Strange Attractor, 2015) Bernie Krause, *The Great Animal Orchestra* (Boston: Back Bay, 2013), and Michael Spitzer, *The Musical Human* (London: Bloomsbury, 2021).

On the strange behavior of noises deep in the sea, Robert Urick, *Principles of Underwater Sound* (New York: McGraw Hill, 3rd ed., 1983 [originally published 1967]), is the classic text that Navy sonar cadets still study today. John Richardson et al., *Marine Mammals and Noise* (San Diego: Academic Press, 1995), is the classic book on the sound pollution issue, and it's also a fine introduction to the bioacoustics of whales themselves. Michael Stocker's *Hear Where We Are* (New York: Springer, 2013) is also a well-informed project.

Since we are so unsure of cetaceans' real lives, we have never been afraid to make things up about them. Here are a few novels where the *sound* of whales takes a central focus. Zakes Mda, *The Whale Caller* (New York: Farrar, Straus and Giroux, 2005), concerns a man who has an easier time talking to whales than to people, until he falls in love. The late Ted Mooney's *Easy Travels to Other Planets* (New York: Farrar, Straus, and Giroux, 1981) presents the difficult choices a woman faces when she learns too much about dolphin sex. Jodi Picoult's *Songs of the Humpback Whale* (Boston: Faber & Faber, 1992) considers the love life of a whale scientist who becomes a hero when he has to save a trapped whale. Witi Imahaera's *The Whale Rider* (New York: Harcourt, 1987) imagines generations of singing whales linking the Maori to the sea. Christopher Moore, *Fluke* (New York: Morrow, 2003), is surely the only book in which a whale telepathically asks us to bring him a hot pastrami on rye. Paul Quarrington's *Whale Music (* New York: Random House, 1997) concerns a rock star hermit locked in his studio composing an electronic symphony for the orcas. Recently I enjoyed Audrey Schulman's *The Dolphin House* (New York: Europa 2022), a fictional account of Margaret Howe's life with Lilly, Bateson, and the dolphin called Peter.

In poetry whales speak in "The Welfleet Whale" by Stanley Kunitz. Colin Teevan wrote "what we said about the whale and what the whale said about us" about a bottlenose whale that got stuck in the Thames a few years ago. D. H. Lawrence wrote "Whales Weep Not," the most famous poem about whale sex, and there are poems by Les Murray, W. S. Merwin, Michael McClure, and countless other humans inspired by the mystery of the cetacean world. Scott McVay's long awaited book of poems *Whales Sing* (Wild River, 2013) is essential reading. The incomparable *Whale Nation* (New York: Harmony, 1988), by the late Heathcote Williams, a book-length poem with photographs and references, is essential reading, a whale epic like no other.

Recent years have brought a few fine whale documentaries that touch on the songs of whales. These include *The Humpback Code*, *The Loneliest Whale*, *Fathom*, and Nat Geo's *Secrets of the Whales.* I still hope there will be more media experiences that get us closer to the whale's music in itself.

Still, nothing you may read or see will be as good as listening for yourself, out on the open sea.

In Memoriam

Pete Seeger

Samone Yust

Joan Ocean McIntyre

James O'Donnell

George Crumb

Ted Mooney

Mira Calix

Vsevolod Bel'kovich

# NOTES

## *One: We Didn't Know, We Didn't Know*

9 **five times faster in water than air:** A. H. Bass and Christopher Clark, "The Physical Acoustics of Underwater Sound Communication," *Springer Handbook of Auditory Research* (New York: Springer Verlag, 2002), p. 19.

10 **"a whale would utter the most doleful groans":** Charles Nordhoff, *Whaling and Fishing* (New York: Dodd and Mead, 1895 [1856]), p. 202.

10 **"it has been known for a long while that humpbacks sing":** H. L. Aldrich, *Arctic Alaska and Siberia* (New York: Rand McNally, 1889).

11 **whales give the ocean its voice:** Roger Payne, *Among Whales* (New York: Scribner's, 1995), p. 145.

13 **If you speed up a humpback whale song it sounds just like a bird:** See https://youtu.be/M5OCCuCIMbA

20 **"whether Leviathan can long endure so wide a chase":** Herman Melville, *Moby Dick* (New York: Modern Library, 1930 [1851]), p. 456.

21 **whales may be gone before we are able to understand them:** Scott McVay, "The Last of the Great Whales," *Scientific American*, vol. 215, no. 2, August 1966, p. 3.

22 **this new sound came in with breathtaking clarity:** Frank Watlington, *How to Build and Use Low-Cost Hydrophones* (Blue Ridge Summit, PA: TAB Books, 1979), p. 109.

25 **"a voice calls us to turn back":** Roger Payne, "The Song of the Whale" [booklet accompanying the original record], *Songs of the Humpback Whale* (Del Mar, CA: Communication Research Machines, 1970), p. 36.

25 **"from the throat of a 40-ton canary":** "Sing, Cetacea, Sing!" *Time*, June 22, 1970, pp. 60–61.

25 **"the inconsequent anguish of an atonal violin":** Albert Goldman, "A Hip New Gift from the Sea," *Life*, vol. 70, no. 16, April 1971, p. 62.

25 **"This is a good record, dig?":** Jon Carroll, "Songs of the Humpback Whale: Review," *Rolling Stone*, 64, August 6, 1970, p. 39.

26 **"the kids were lacing their whale songs with pot":** Joseph Morgenstern, "Whale Songs," *Newsweek*, February 15, 1971, p. 16.

26 **McVay embarked upon a remarkable journey to Japan:** This trip is eloquently summarized in his unpublished report entitled "Initiation of Whale Campaign in Japan" (New York: Environmental Defense Fund, August 1970).

28 **"I didn't remember a thing about whales myself":** Kenzaburo Oe, from "The Day the Whale Becomes Extinct," trans. John Nathan. Quoted in Scott McVay, "What on Earth Was a Whale?" *Animal Welfare Institute Information Report*, vol. 23, no. 1, spring 1974, p. 3.

30 **"as unthinkable as taking all music away":** Scott McVay, "Can Leviathan Endure So Wide a Chase?" *Natural History*, vol. 80, no.1, January 1971, p. 72.

30 **"sold more than thirty million copies":** For those who are interested, the highest-selling record of all time is reputed to be Michael Jackson's *Thriller*, clocking in at over sixty million copies sold.

## *Two: Gonna Grow Fins*

35 **"In music, 1970 was the year of the whale":** Harold Schonberg, "Of Humpback Whales, Those Seven Veils, and Other Tales," *New York Times*, January 3, 1971, p. D13.

36 **The whales *do* sing Oriental music:** Bruce Duffie's 1985 interview with Alan Hovhaness is at http://www.bruceduffie.com/hovx.html.

36 **"Hovhaness beamed like an ecologist":** "Sing, Cetacea, Sing!" *Time*, June 22, 1970, p. 59.

36 **"Mr. Hovhaness must have suspected he would be harpooned":** Donal Henehan, "Whales Sing Out," *New York Times*, June 13, 1970, p. 21.

36 **the one piece he regretted having written:** Bruce Duffie's 1985 interview with Alan Hovhaness, http://www.bruceduffie.com/hovx.html.

38 **"more like a mewling cat than a nightingale":** Paul Kresh, "Whales Can't Sing, Silly!" *Stereo Review*, no. 25, August 1970, p. 100.

43 **"'Vox Balaenae' may not turn out to be a major work:** Donal Henehan, "Inventive 'Voice of the Whale,'" *New York Times*, April 7, 1973, p. 40.

44 **"Can you see a *whale* as a rock star *heh heh*":** Roger Ames, interview with Captain Beefheart (Don Van Vliet) [c. 1974] at http://www.beefheart.com/captain-beefheart-by-roger-ames/.

45 **"We heard our screams turn into songs":** Lou Reed, quoted in Victor Bockris, *Transformer: The Lou Reed Story* (New York: Simon & Schuster, 1994), p. 96.

46 **"meditation and water are wedded for ever":** Herman Melville, *Moby Dick*, p. 2.

53 **"Cock-a-doodle-do produces music too":** From the U.S. House Subcommittee on International Organizations and Movements, July 26, 1971, quoted in Lewis Regenstein, *The Politics of Extinction* (New York: Macmillan, 1975), p. 69.

54 **"No permit may be issued for the taking of any marine mammal":** The U.S. Marine Mammal Protection Act of 1972 may be found at https://www.fisheries.noaa.gov/national/marine-mammal-protection/marine-mammal-protection-act.

54 **The elegant book drew further support for these wonderful animals:** Joan McIntyre, ed., *Mind in the Waters* (New York: Scribner's, 1974).

57 **I got to perform it together with him, one final tribute:** See *Pete Seeger: Storm King*, https://www.audible.com/pd/Pete-Seeger-Storm-King-Volume-2-Audiobook/B01A9BHQME

## *Three: Those Orcas Love a Groove*

61 **Now available cheaply from China:** You can get reasonable underwater speakers from Daravoc/Esunpride: https://www.ebay.com/itm/273006179803.

62 **"I get no response from you and I'm getting *bugged*":** Paul Horn, *Inside Paul Horn* (San Francisco: HarperCollins, 1990), p. 179.

64 **the whales were conducting their own experiments on him:** Erich Hoyt, *Orca: The Whale Called Killer* (Tonawanda, NY: Firefly Books, 1990), p. 43.

65 **"Sometimes they seem to join in the celebration":** Paul Spong, "The Whale Show," in McIntyre, *Mind in the Waters*, p. 180.

65 **"I've just walked into the opera house, I have no program":** Hoyt, *Orca*, p. 53.

66 **We're not going to lose that song now:** I got to record it together with him, one final tribute, on *Pete Seeger: Storm King*, https://www.audible.com/pd/Pete-Seeger-Storm-King-Volume-2-Audiobook/B01A9BHQME <https://www.audible.com/pd/Pete-Seeger-Storm-King-Volume-2-Audiobook/B01A9BHQME>

66 **"the charged border as a luminous crack between worldviews":** Jim Nollman, *The Charged Border: Where Whales and Humans Meet* (New York: Henry Holt, 1999), p. 62.

68 **"the difference between humans and angels":** Nollman, *The Charged Border*, p. 222.

74 **"each of us became all one hollowed and listening ear":** James Agee and Walker Evans, *Let Us Now Praise Famous Men* (Boston: Houghton Mifflin, 1941), p. 463.

74–75 **"in perfect irony and silence withdrawn":** Agee and Evans, *Let Us Now Praise Famous Men*, p. 470.

76 **it's called N47:** John K. B. Ford, "Acoustic behaviour of resident killer whales off Vancouver Island, British Columbia," *Canadian Journal of Zoology*, vol. 67, 1989, pp. 727–745.

76 **"it was enough to make me want to study mountain beavers instead":** Bruce Obee and Graeme Ellis, *Guardians of the Whales* (Anchorage: Alaska Northwest Books, 1992), p. 30.

77 **showing in fascinating detail how the population has changed:** John K. B. Ford, Graeme Ellis, and Kenneth Balcomb, *Killer Whales: The Natural History and Genealogy of Orcinus orca in British Columbia and Washington State* (Seattle: University of Washington Press, 2nd ed., 2000).

77 **orca calls help to coordinate activities:** Ford, "Acoustic behaviour of resident killer whales," p. 727.

77 **certain orca calls are within pod, others are from pod to pod:** F. Thomsen, V. Deecke, and John K. B. Ford, "Characteristics of whistles from the acoustic repertoire of resident killer whales off Vancouver Island, British Columbia," *Journal of the Acoustical Society of America*, vol. 109, no. 3, March 2001, pp. 1240–1246.

77 **"certain family-specific calls are much more frequent following the birth of a calf":** Bridget Weiss, Friedrich Ladich, Paul Spong, and Helen Symonds, "Vocal behavior of resident killer whale matrilines with newborn calves: The role of family signatures," *Journal of the Acoustical Society of America*, vol. 119, no. 1, January 2006, pp. 627–635.

77 **killer whales have feelings too:** Rüdiger Riesch, John K. B. Ford, and Frank Thomsen, "Stability and group specificity of stereotyped whistles in resident killer whales off British Columbia," *Animal Behaviour*, vol. 71, 2006, pp. 79–91.

78 **in New Zealand the pods are nowhere near as distinct:** Ingrid Visser, *Swimming with Orca* (Auckland: Penguin NZ, 2006), p. 114.

79 **certain parts of the sound change without rhyme or reason:** Volcker Deecke, John K. B. Ford, and Paul Spong, "Dialect change in resident killer whales: implications for vocal learning and cultural transmission," *Animal Behaviour*, vol. 60, 2000, pp. 629–638.

81 **"He would show us his flukes one last time and be gone":** Kenneth Brower, *Freeing Keiko: The Journey of a Killer Whale from* Free Willy *to the Wild* (New York: Gotham Books, 2005), p. 300.

81 **a woman was thrown in jail for patting Luna on the head:** Michael Parfit, "A Whale of a Tale," *Smithsonian*, vol. 35, no. 8, November 2004, p. 67.

84 **"Sometimes I had to close my eyes to concentrate":** Jim Nollman, "Sex, Dolphins, and Rock 'n' Roll" [1994] https://www.whaleweb.org/intersp/pages/SEX_DOL.html .

*Four: To Hear the Dolphin Call His Name*

90 **"The extraterrestrials are already here—in the sea":** John Lilly, *The Mind of the Dolphin* (New York: Doubleday, 1967), p. 40.

90 **"The dim outlines of a Someone began to appear":** John Lilly, "A Feeling of Weirdness," in McIntyre, *Mind in the Waters*, p. 71.

91 **"he will never leave your side":** Lilly, *The Mind of the Dolphin*, p. 266.

91 **"this living is taxing on my private life":** Lilly, *The Mind of the Dolphin*, p. 273.

92 **"I, for some reason, left out things about myself":** Lilly, *The Mind of the Dolphin*, pp. 282–283.

94 **"It's going to be about relationships":** Gregory Bateson, "Problems in Cetacean and Other Mammalian Communication," in *Steps to an Ecology of Mind* (New York: Bantam, 1977), p. 370.

96 **it is sometimes called "acoustic tissue":** T. W. Cranford, "In search of impulse sound sources in odontocetes," in Whitlow Au et al., ed., *Hearing by Whales and Dolphins* (Berlin: Springer Verlag, 2000), pp. 109–155.

96 **"to shape the beam of sound":** Sylvia van der Woude, "Effects of Signature Whistle Playbacks on Bottlenose Dolphins" (unpublished master's thesis, Freie Universität Berlin, 2003), p. 30.

98 **dolphins may store "sound pictures" in their brains:** Adam Pack, Louis Herman, Matthias Hofmann-Kuhnt, and Brian Branstetter, "The object behind the echo: dolphins perceive object shape globally through echolocation," *Behavioural Processes*, vol. 58, 2002, pp. 1–26.

99 **a particular sound unique to each individual:** Melba Caldwell and David Caldwell, "Individualized whistle contours in bottlenosed dolphins," *Nature*, no. 207, 1965, pp. 434–435.

99 **"vocal learning allows a remarkable open-ended system of communication":** Peter Tyack, "Cetaceans: a comparative and evolutionary framework," in Herbert Roitblat, Louis Herman, and Paul Nachti-gall, eds. *Language and Communication: Comparative Perspectives* (Hillsdale, NJ: Lawrence Erlbaum Associates, 1993), p. 131.

100 **she imitated no other animal:** Tyack, "Cetaceans," p. 137.

100 **we know little about how these sounds function in these animals' usual life:** Vincent Janik and Peter Slater, "Context-specific use suggests that bottlenose dolphin signature whistles are cohesion calls," *Animal Behaviour*, vol. 56, 1998, pp. 829–838.

101 **dolphin mothers recognize the sounds of their adult children:** Laela Sayigh, Peter Tyack et al., "Individual recognition in wild bottlenose dolphins: a field test using playback experiments," *Animal Behaviour*, vol. 57, 1998, pp. 41–50.

101 **lone captive dolphins produce a shared whistle most of the time:** Brenda McCowan and Diana Reiss, "The fallacy of 'signature whistles' in bottlenose dolphins," *Animal Behaviour*, vol. 62, 2001, pp. 1151–1162.

102 **it is the contour not the tone of the whistle that matters:** Vincent Janik, Laela Sayigh, and Randall Wells, "Signature whistle shape conveys identity information to bottlenose dolphins," *Proceedings of the National Academy of Sciences*, vol. 103, no. 21, 2006, pp. 8293–8297.

103 **McCowan and Reiss used a classification method that is biased:** Laela Sayigh, Carter Esch, Randall Wells, and Vincent Janik, "Facts about signature whistles of bottlenose dolphins," *Animal Behaviour*, vol. 74, 2007, doi:10.1016j.anbehav.2007.02.018.

104 **In Silbo, five vowels and four consonants are produced:** R. G. Busnel and A. Classe, *Whistled Languages* (Berlin: Springer Verlag, 1976), p. 53. See also Julien Meyer, Marcelo O. Magnasco, and Diana Reiss, "The Relevance of Human Whistled Languages for the Analysis and Decoding of Dolphin Communication," *Frontiers in Psychology*, vol. 12, 2021, pp. 9–30.

105 **the most frequently used signal is a contact call that conveys signature information about the caller:** Diana Reiss, *The Dolphin in the Mirror*, Houghton Mifflin Harcourt, 2011, p. 106.

106 **"a rational animal can perceive how its world is structured":** Louis Herman, "Intelligence and rational behaviour in the bottlenosed dolphin," in Susan Hurley and Matthew Nudds, eds., *Rational Animals?* (London: Oxford University Press, 2006), pp. 439–467.

107 **dolphins can stun their prey with loud and sudden sounds:** Eugenie Samuel, "Killer Clicks," *New Scientist*, January 31, 2001. https://www.newscientist.com/article/dn371-killer-clicks/

107 **"do you want to keep your myth?":** Melba Caldwell, quoted in Michael Parfit, "Are dolphins trying to say something, or is it all much ado about nothing?" *Smithsonian*, vol. 11, October 1980, p. 80.

108 **"I was pegged as a dolphin man":** Buck Henry, speaking on the commentary on the DVD of *The Day of the Dolphin*, dir. Mike Nichols (Los Angeles: Image Entertainment 2006 [original release date 1973]). See also Peter Feibleman, "Mike Nichols Tries to Make a 'Talkie' with Dolphins," *Atlantic*, vol. 233, January 1974, pp. 71–81.

109 **This was called swimmer nullification:** David Helvarg, *Blue Frontier* (San Francisco: Sierra Club Books, 2nd ed., 2006), p. 71.

110 **"we don't need to share war with our fellow mammals":** Helvarg, *Blue Frontier*, p. 73.

110 **dolphins never wage war:** John Lilly, *Man and Dolphin* (New York: Pyramid Books, 1962).

110 **"the wastes of water beyond the power of man to describe":** Loren Eiseley, "Man and Porpoise: Two Solitary Destinies," *American Scholar*, vol. 30, no. 1, Winter 1960–61, p. 60.

## *Five: Beluga Do Not Believe in Tears*

115 **"they are said to lament when they are caught":** Adriaen Coenen, *The Whale Book* (London: Reaktion Books, 2003 [1585]), p. 90.

116 **former dolphin trainers now releasing animals:** Richard O'Barry, *To Free a Dolphin* (Los Angeles: Renaissance Books, 2000).

117 **a seven-hundred-page book on incentive training:** Ken Ramirez, ed., *Animal Training: Successful Animal Management Through Positive Reinforcement* (Chicago: Shedd Aquarium, 1999).

122 **the basic sounds made by captive belugas:** Cheri Ann Recchia, "Social Behaviour of Captive Belugas" (unpublished Ph.D. dissertation, Massachusetts Institute of Technology/Woods Hole Oceanographic Institution Joint Program in Oceanography, 1994).

122 **an exhaustive survey of the vocal repertoire of wild belugas:** B. L. Sjare and T. G. Smith, "The vocal repertoire of white whales summering in Cunningham Inlet, Northwest Territories," *Canadian Journal of Zoology*, vol. 64, 1986, pp. 407–415.

122 **belugas may have signature vowels:** Elena Panova et al., "Intraspecific variability in the 'vowel'-like sounds of beluga whales (*Delphinapterus leucas*)," *Marine Mammal Science*, vol. 32, no. 2, 2016, pp. 452–465.

124 **"I brought Nollman here and we made this rubbing":** Jim Nollman, "Talking to Beluga," in Toni Frohoff and Brenda Peterson, ed., *Between Species: Celebrating the Dolphin-Human Bond* (San Francisco: Sierra Club Books, 2003), pp. 91–103.

125 **"they call whales to the beaches with powerful charms":** Adam of Bremen, *History of the Archbishops of Hamburg-Bremen*, trans. Francis Tschan (New York: Columbia University Press, 2002 [1085]), p. 212.

125 **two hundred thousand people died:** Anne Applebaum, *Gulag* (New York: Doubleday, 2003), p. 65. See also Cynthia Ruder, *Making History for Stalin: The Story of the Belomor Canal* (Gainesville: University Press of Flórida, 1998).

126 **the woman who fell in love with a whale:** this Chukchi story appears in Yuri Rythkheu, *A Dream in Polar Fog*, trans. Ilona Chavasse (Brooklyn, NY: Archipelago Books, 2005), pp. 104–107.

127 **"Then was there heard a most celestial sound":** Edmund Spenser, "Amoretti," canto 38 [1594], in Francis J. Child, ed., *The Poetical Works of Edmund Spenser.* http://www.gutenberg.org/etext/10602.

130 **several papers from the Shirshov Institute in English:** Karen Pryor and Kenneth Norris, eds., *Dolphin Societies* (Berkeley: University of California Press, 1991).

131 **"the Gulag Archipelago":** In Solzhenitsyn's book of the same name, this phrase refers to the entire archipelago of prison camps throughout the Soviet Union, as well as this one particular island group where the system was administered from.

131 **"not so much a color as the visible absence of color":** Herman Melville, *Moby Dick*, p. 282.

132 **"Listen with you ears, not your ass":** Matti Kuusi, *Proverbia Septentrionalia* (Helsinki: Academia Scientiarum Fennica, 1985), pp. 328–329.

136 **"arriving across the sea accompanied by two white whales":** Hernan Patiño and Jöns Carlson, *Maitovalat: Vinenanmeren luostarisaaren* (Helsinki: Maahenki Oy, 2003).

138 **Was this whale a Communist spy?:** The Turkish beluga story is recounted by Pierre Béland, *Beluga: A Farewell to Whales* (New York: Lyons and Burford, 1996), pp. 113–163.

140 **plenty of chirping during beluga sex, but absolutely no bleating:** R. A. Belikov and V. M. Bel'kovich, "Underwater Vocalization of the Beluga Whales in a Reproductive Gathering in Various Behavioral Situations," *Oceanology*, vol. 43, no. 1, 2003, pp. 112–120.

140 **belugas may have signature vowels:** V. M. Bel'kovich and S. A. Kreichi, "Specific Features of Vowel-Like Signals of White Whales," *Acoustical Physics*, vol. 50, no. 3, 2004, pp. 288–294. See also V. M. Bel'kovich and M. N. Sh'ekotov, *The Belukha Whale: Natural Behavior and Bioacoustics*, trans. Marina Svanidze (Woods Hole, MA: Woods Hole Oceanographic Institution, 1993). https://darchive.mblwhoilibrary.org/handle/1912/75.

142 **"A piece of whale in each man's body":** Elias Lönnrot, *Suomen Kansan Vanhat Runot* [The Ancient Poems of the Finnish People], ed. Matti Kuusi and Senni Timonen (Helsinki: Suomalaisen Kirjallisuuden Seura, 1997), vol. 15, poem 530.

## *Six: The Longest Liquid Song*

147 **the honeybee represents complex information:** Karl von Frisch, *The Dance Language and Orientation of Bees* (Cambridge: Harvard University Press, 1967). See also "Waggle dance leads bees to nectar," *BBC News*, May 11, 2005. http://news.bbc.co.uk/1/hi/sci/tech/4536127.stm.

147 **bowerbirds function sculptures out of trees and leaves:** Jared Diamond, "Evolution of Bowerbirds' Bowers: Animal Origins of the Esthetic Sense," *Nature*, vol. 297, 1982, p. 99. See also Clifford Frith et al., *The Bowerbirds* (London: Oxford University Press, 2004) and A. J. Marshall, *Bowerbirds: Their Displays and Breeding Cycles* (London: Oxford University Press, 1954).

148 **A single humpback performance can last twenty-three hours:** Roger Payne, *Among Whales*, p. 144.

148 **"a series of surprisingly beautiful sounds":** Roger Payne and Scott McVay, "Songs of Humpback Whales," *Science*, vol. 173, no. 3997, August 13, 1971, p. 585.

151 **"the whole ordered series of themes is called a song":** Most researchers still follow the same basic terminology proposed by Payne and McVay in 1971, although some researchers contend that the category of a whole song may be arbitrary because the beginnings and ends are only when the whale needs to breathe. See Daniela Cholewiak, Renate Sousa-Lima, and Salvatore Cerchio, "Humpback whale song hierarchical structure: Historical context and discussion of current classification issues," *Marine Mammal Science*, vol. 29, no. 3, 2013, pp. 312–332.

151 **on birds with complex songs:** David Rothenberg, *Why Birds Sing* (New York: Basic Books, 2005).

155 **humpback songs can be said to rhyme:** Linda Guinee and Katharine Payne, "Rhyme-like Repetitions in Songs of Humpback Whales," *Ethology*, vol. 79, 1988, pp. 295–306. (This research was openly discussed in the early 1970s even though it took more than a decade to publish the paper.)

155 **the song is only performed by males:** H. E. Winn, W. L. Bischoff, and A. G. Taruski, "Cytological sexing of cetaceans," *Marine Biology*, vol. 23, 1972, pp. 343–346.

156 **songs lengthen from year to year:** Katherine Payne, Peter Tyack, and Roger Payne, "Progressive Changes in the Songs of Humpback Whales: A Detailed Analysis of Two Seasons in Hawaii," in Roger Payne, ed., *Communication and Behavior in Whales* (Boulder, CO: Westview Press, 1983), pp. 9–57.

158 **"today's humpback whale songs pale beside those of the sixties":** Roger Payne, "An Open Letter to the Youth of Japan" (2005) https://whale.org/sympathetic-voice-open-letter-japan-dr-roger-payne/.

158 **on the yellow-rumped cacique and the village indigobird:** Robert Payne, "Behavioral continuity and change in local song populations of village indigobirds," *Zeitschrift für Tierpsychologie*, vol. 70, 1985, pp. 1–44. J. M. Trainer, "Cultural evolution in song dialects of yellow-rumped caciques in Panama," *Ethology*, vol. 80, 1989, pp. 190–204.

159 **singing the same song, separated by thousands of ocean miles:** Salvatore Cerchio, Jeff Jacobson, and Thomas Norris, "Temporal and geographical variation in songs of humpback whales: synchronous change in Hawaiian and Mexican breeding assemblages," *Animal Behaviour*, vol. 62, 2001, pp. 313–329.

161 **he heard moans, sighs, and chainsaws:** Douglas Cato, "Songs of humpback whales: the Australian perspective," *Memoirs of the Queensland Museum*, vol. 30, no. 2, 1991, pp. 277–290.

161 **she called the final sound a *screal*:** Astrid Mednis, "An acoustic analysis of the 1988 song of the humpback whale off Eastern Australia," *Memoirs of the Queensland Museum*, vol. 30, no. 2, pp. 323–332.

162 **cultural evolution was not supposed to happen that fast:** Michael Noad et al., "Cultural revolution in whale songs," *Nature*, no. 408, 2000, p. 537.

163 **they heard nine basic kinds of sounds:** Whitlow Au et al., "Acoustic properties of humpback whale songs," *Journal of the Acoustical Society of America*, vol. 120, no. 2, August 2006, pp. 1103–1110.

166 **each part of the song might have a different function:** Eduardo Mercado, Louis Herman, and Adam Pack, "Stereotypical sound patterns in humpback whale song," *Aquatic Mammals*, vol. 29, no. 1, 2003, pp. 37–52. See also Sean Green, Eduardo Mercado, Louis Herman, and Adam Pack, "Characterizing Patterns within Humpback Whale Songs," *Aquatic Mammals*, vol. 33, no. 2, 2007, pp. 202–213, and Louis Herman, "The multiple functions of male song within the humpback whale mating system: review, evaluation, and synthesis," *Biological Reviews*, vol. 92, no. 3, 2017, pp. 1795–1818.

166 **could humpback whale song be sonar?:** L. N. Frazer and E. Mercado, "A Sonar Model for Humpback Whale Song," *Journal of Oceanic Engineering*, vol. 25, no. 1, January 2000, pp. 160–181. See also Whitlow W. L. Au, Adam Frankel, David A. Helweg, and Douglas H. Cato, "Against the Humpback Whale Sonar Hypothesis," *Journal of Oceanic Engineering*, vol. 26, no. 2, April 2001, pp. 295–301.

166 **could it be true that only the males need sonar?:** Eduardo Mercado and Christina E. Perazio, "Similarities in Composition and Transformations of Songs by Humpback Whales Over Time and Space," *Journal of Comparative Psychology*, vol. 135, no. 1, 2021, pp. 28–50. See also Eduardo Mercado, "Song Morphing by Humpback Whales: Cultural or Epiphenomenal?" *Frontiers in Psychology*, January 2021, https://doi.org/10.3389/fpsyg.2020.574403

166 **how long a male could hold his breath under water:** N. C. Chu and P. Harcourt, "Behavioural correlations with aberrant patterns in humpback whale songs," *Behavioural Ecology and Sociobiology*, no. 19, 1986, pp. 309–312.

166 **could the cry be a humpback signature whistle?:** George Hafner et al., "Signature information in the song of the humpback whale," *Journal of the Acoustical Society of America*, vol. 66, no. 1, July 1979, pp. 1.

167 **she found innovation in both directions:** Nina Eriksen, Lee Miller, Jakob Tougaard, and David Helweg, "Cultural change in the songs of humpback whales from Tonga," *Behaviour*, no. 142, 2005, pp. 305–328.

170 **White-crowned sparrows lengthened their songs:** Elizabeth Derryberry, "Evolution of bird song affects signal efficacy," *Evolution*, vol. 61, no. 8, August 2007, pp. 1938–1945.

170 **birds have a natural aesthetic sense:** Charles Darwin, *The Descent of Man* (Chicago: Brittanica, 1952 [1871]), p. 451.

171 **whale songs lack information:** Ryuji Suzuki, John Buck, and Peter Tyack, "Information entropy of humpback whale songs," *Journal of the Acoustical Society of America*, vol. 119, no. 3, March 2006, pp. 1849–1866.

172 **music meant more to our ancestors than it could ever mean for us:** Steven Mithen, *The Singing Neanderthals* (Cambridge: Harvard University Press, 2006).

173 **whales may experience complex emotion:** Andy Coghlan, "Whales boast the brain cells that make us human," *New Scientist*, November 27, 2006. See also Patrick Hof and Estel van der Gucht, "Structure of the cerebral cortex of the humpback whale," *Anatomical Record Part A*, 290, 2007, pp. 1–31, and Lori Marino et al., "Cetaceans have complex brains for complex cognition," *PLOS Biology*, vol. 5, no. 5, 2007, e139. doi:10.1371/journal.pbio.0050139.

174 **It may depend on emotion, beyond any reason:** See Anita Murray,Rebecca A. Dunlop, and Michael J. Noad, "Stereotypic and complex phrase types provide structural evidence for a multi-message display in humpback whales," *Journal of the Acoustical Society of America*, 2018, pp. 980–994. See also Daniela Cholewiak, Renate Sousa-Lima, and Salvatore Cerchio, "Humpback whale song hierarchical structure: Historical context and discussion of current classification issues," *Marine Mammal Science*, vol. 29, no. 3, 2013, pp. 312–332.

176 **humpback whales as giant underwater submarines:** Christopher Moore, *Fluke* (New York: William Morrow, 2003).

176 **song tended to temporarily gather males:** James Darling and Martine Bérubé, "Interactions of singing humpback whales with other males," *Marine Mammal Science*, vol. 17, no. 3, July 2001, pp. 570–584.

177 **167 singing whale/other whale encounters:** James Darling, Meagan Jones, and Charles Nicklin, "Humpback whale songs: Do they organize males during the breeding season?" *Behaviour*, no. 143, 2006, pp. 1051–1101.

## *Seven: Moby Click*

185 **"the mode by which this is effected remains a curious secret":** Thomas Beale, *A Few Observations on the Natural History of the Sperm Whale* (London: Effingham Wilson, 1835), ch. 2, §5. https://whalesite.org/anthology/bealeold.htm.

186 **the sperm whale is an animal of extremes:** Hal Whitehead, *Sperm Whales: Social Evolution in the Ocean* (Chicago: University of Chicago Press, 2003), p. 2.

186 **tracking sperm whales aboard the sloop:** *Tulip*: Hal Whitehead, *Voyage to the Whales* (White River Junction, VT: Chelsea Green, 1990).

192 **the argument for culture in cetaceans is quite simple:** Luke Rendell and Hal Whitehead, "Culture in Whales and Dolphins," *Behavioral and Brain Sciences*, vol. 24, 2001, pp. 309–382.

192 **humpbacks in Brazil stick their tails in the air while they sing:** Renata Sousa-Lima, "Acoustic Ecology of Humpback Whales in the Abrolhos National Marine Park, Brazil" (unpublished Ph.D. dissertation, Cornell University, Ithaca, NY, 2007).

193 **animal culture could be called "normal" culture:** Jerome Barkow, "Why can't a cetacean be more like a (hu)man?" *Behavioral and Brain Sciences*, vol. 24, 2001, p. 325.

193 **doubt that cetaceans share meaning in a cognitive way:** Robin Dunbar, "So how *do* they do it?" *Behavioral and Brain Sciences*, vol. 24, 2001, p. 332.

189 **parrots certainly possess social learning ability:** Spencer Lynn and Irene Pepperberg, "Culture: In the beak of the beholder?" *Behavioral and Brain Sciences*, vol. 24, 2001, p. 341.

189 **why do some species learn social behavior?:** Peter Slater, "There's culture and *Culture*," *Behavioral and Brain Sciences*, vol. 24, 2001, p. 356.

189 **which differences are trivial and which are not:** David Premack and Marc Hauser, "Calling it culture doesn't help," *Behavioral and Brain Sciences*, vol. 24, 2001, p. 350.

190 **what scientists heard as cacophony made sense to the drum master:** Michel André and Cees Kamminga, "Rhythmic dimension in the echolocation click trains of sperm whales," *Journal of the Marine Biological Association of the United Kingdom*, vol. 80, 2000, pp. 163–169.

192 **a more mathematically rigorous way to categorize clicks:** Mike van der Schaar and Michel André, "An Alternative Sperm Whale Coda Click Naming Protocol," *Aquatic Mammals*, vol. 32, no. 3, 2006, pp. 370–373.

193 **"the ordering of codas is to some extent nonrandom":** Linda Weilgart, "Vocalizations of the sperm whale off the Galapagos Islands as related to behavioral and circumstantial variables" (unpublished Ph.D. dissertation, Dalhousie University, Halifax, 1990), p. 109.

193 **computers have got to have more Africa in them:** Brian Eno, quoted in Kevin Kelly, "Gossip is Philosophy," *Wired*, vol. 3, no. 5, May 1995, https://www.wired.com/1995/05/eno-2/.

195 **"the fishing method seems to be initiated by the dolphins":** Karen Pryor, Scott Lindbergh, and Raquel Milano, "A Dolphin-Human Fishing Cooperative in Brazil," *Marine Mammal Science*, vol. 6, no. 1, January 1990, pp. 77–82.

195 **together orca and human would make the kill:** Danielle Clode, *Killers in Eden* (Sydney: Allen and Unwin, 2002). See also https://killerwhalemuseum.com.au/meet-the-killers/.

## *Eight: Thousand Mile Song*

203 **they came up with 1,435 meters per second:** Victoria Kahari, "Sounding Out the Ocean's Secrets" (Washington, DC: National Academy of Sciences, 2003). http://www.beyonddiscovery.org/content/view.article.asp?a=219.

205 **the sound waves vibrate and travel endlessly:** Robert Urick, *Principles of Underwater Sound* (New York: McGraw Hill, 3rd ed., 1983 [1967]), p. 159.

212 **"by bouncing their sounds off the ice, the whales may reconstruct an image of their underwater world":** Chris Clark speaking on "Acoustic Herd," produced by Sherre de Lys and John Jacobs for *Radio Eye* (Sydney: Australian Broadcasting Company, 2000).

213 **still skeptical that such tones could be of animal origin:** Urick, *Principles of Underwater Sound*, p. 218.

215 **these largest whales could communicate across entire oceans:** Payne, *Among Whales*, p. 383.

215 **His latest creation is Dolphin Communicator:** Software for live two-way audio communication with dolphins: https://gitlab.com/serge_masse/android-dolphin-comm-doc.

216 **able to track a blue whale for forty-three days from a thousand miles away:** Chris Clark and William Ellison, "Potential use of low-frequency sounds by baleen whales for probing the environment," in J. A. Thomas, C. F. Moss, and M. Vater, eds., *Echolocation in Bats and Dolphins* (Chicago: University of Chicago Press, 2004), p. 604.

217 **blue whales make three kinds of sounds:** Erin Oleson, Sean Wiggins and John Hildebrand, "Temporal separation of blue whale call types on a Southern California feeding ground," *Animal Behaviour*, vol. 74, no. 4, 2007, pp. 881–894.

219 **blue whale populations can be distinguished by dialect:** Mark McDonald, Sarah Resnick, and John Hildebrand, "Biogeographic characterisation of blue whale song worldwide: using song to identify populations," *Journal of Cetacean Research Management*, vol. 8, no. 1, 2006, pp. 55–65.

221 **"So far this remains an unpublished hypothesis":** Personal conversation with Erin Oleson, August 2007.

222 **songbirds living in cities make louder sounds:** Catherine Brahic, "Urban songbirds raise their voices to be heard," *New Scientist*, December 4, 2006. https://www.newscientist.com/article/dn10720-urban-songbirds-raise-their-voices-to-be-heard/.

223 **"airspace resonance phenomena resulted in hemorrhaging":** Ken Balcomb, letter to the U.S. Navy Low Frequency Sonar Program, February 21, 2001. http://www.oceanmammalinst.com/kenbalcombletter.htm.

223 **more than a hundred whales are known to have died:** Michael Jasny et al., *Sounding the Depths II: The Rising Toll of Sonar, Shipping, and Industrial Ocean Noise on Marine Life* (New York: Natural Resources Defense Council, 2005). https://www.nrdc.org/issues/protect-marine-mammals-ocean-noise.

224 **"we need to track the submarines of our possible enemies":** Tony Perry, "Navy agrees to sonar precautions," *Los Angeles Times*, July 8, 2006.

224 **"a significant step forward":** Perry, "Navy Agrees."

226 **"our actions ripple and interact in ways we cannot predict":** Jim Cummings, "Listening in the Sea," 2006, http://acousticecology.org/writings/listeninginsea.html.

## *Nine: Never Satisfied*

232 **the male whales sometimes charge toward the boat:** Peter Tyack, "Differential response of humpback whales to playback of song or social sounds," *Behavioral Ecology and Sociobiology*, no. 13, 1983, pp. 49–55.

233 **no whale approached a silent boat:** J. R. Mobley, L. M. Herman, and A. S. Frankel, "Responses of wintering humpback whales to playbacks or recordings of winter and summer vocalizations and of synthetic sound," *Behavioral Ecology and Sociobiology*, no. 23, 1988, pp. 211–223.

233 **Humphrey found his way out to sea:** Steve Rubenstein, "Humphrey Remembered," *San Francisco Chronicle*, May 17, 2007.

233 **playback is no sure way to move a whale:** Eric Bailey, "Next for whales: sweet sounds or scary music?" *Los Angeles Times*, May 23, 2007. See also Henry Lee and Glen Martin, "Whales disappear—rescuers believe they're back at sea," *San Francisco Chronicle*, May 30, 2007.

234 **how whales are moving when they make sound:** Mark Johnson and Peter Tyack, "A Digital Acoustic Recording Tag for Measuring the Response of Wild Marine Mammals to Sound," *IEEE Journal of Oceanic Engineering*, vol. 28, no. 1, 2003, pp. 3–12.

235 **those right whales come to the surface after hearing a loud alarm sound:** Paul Rincon, "Whales drawn to emergency sirens," *BBC News*, December 3, 2003. See also Randall Reeves et al., "Report of the North Atlantic Right Whale Program Review," Washington, DC: Marine Mammal Commission, 2007.

242 **he was inspired to resurrect a Maori tale:** Witi Ihimaera, *The Whale Rider* (New York: Harcourt, 1987), p. i.

242 **one Maori myth on whales centers on revenge:** "Kae's Theft of the Whale." http://www.sacred-texts.com/pac/grey/grey07.htm.

257 **"the real music comes when you have no idea what is going on":** Mark Kidel, *Joe Zawinul: A Musical Portrait* [DVD] (London: Calliope Media, 2005). Available here online: https://fb.watch/emKsi2Ej2b/

259 **"what we call music is not the real music":** Evan Parker, "Why 211?" in John Zorn, ed., *Arcana II: Musicians on Music* (New York: Hips Road, 2007), p. 203.

259 **the pattern tells me there may be some sense in it after all:** David Rothenberg, "To Wail with a Whale: Anatomy of an Interspecies Duet," *TRANS: The Journal of the Spanish Society of Ethnomusicology*, 12, 2008. https://www.sibetrans.com/trans/article/97/to-wail-with-a-whale-anatomy-of-an-interspecies-duet. This article describes my duet with the humpback whale in much greater detail than in this book.

260 **Iceland ends commercial whaling:** Charles Clover, "Iceland ends whaling due to lack of demand," *Daily Telegraph*, August 24, 2007.

## *Ten: Whalesong Inside Us*

266 **"What is this creature able to do and be?"** Martha Nussbaum, "What we owe our fellow animals," *New York Review of Books*, Mar. 10, 2022, issue, https://www.nybooks.com/articles/2022/03/10/what-we-owe-our-fellow-animals-ethics-martha-nussbaum/

268 **That is why we developed unique glyphs:** David Rothenberg and Michael Deal, "A New Morphological Notation for the Music of Humpback Whales," *Art and Perception*, 2015. See also https://medium.com/@dealville/whales-synchronize-their-songs-across-oceans-and-theres-sheet-music-to-prove-it-b1667f603844

278 **They hope to have the first results by 2026:** Jakob Andreas et al., "Toward understanding the communication in sperma whales," *iScience* 25, 104393, June 17, 2022. https://www.cell.com/iscience/fulltext/S2589-0042(22)00664-2

285 **the film *The Loneliest Whale*, directed by Josh Zeman:** https://bleeckerstreetmedia.com/the-loneliest-whale

286 **an ambisonic surround-sound installation for the renovated Whale Hall:** https://www.terrain.org/2020/currents/the-moving-sounds-of-whales/

294 **$1 trillion for the current stock of great whales:** Ralph Chami et al., "Nature's Solution to Climate Change," *IMF Finance and Development*, December 2019, p. 36. https://www.imf.org/en/Publications/fandd/issues/2019/12/natures-solution-to-climate-change-chami

313 **No air leaves the whale while he sings:** Juliette Damien, Olivier Adam, et al, "Anatomy and morphology of the Mysticete rorqual larynx," *The Philosophical Record*, 302, 2018, pp. 703-717. See also Dorian Cazao, Olivier Adam, et al, "A study of vocal nonlinearities in humpback whale songs: from production mechanisms to acoustic analysis," *Scientific Reports* 6, 31660, 2018, pp. 1-12.

# ILLUSTRATION CREDITS

Most two-page chapter opener photos by Wade and Robyn Hughes, from their book *Looking for Whales* (Halstead Press, 2019).

**Wade Hughes** has repeatedly been an *Australian Geographic* Nature Photographer of the Year finalist, winning the 2016 award for the year's best animal portrait. In 2017 Britain's Natural History Museum selected his humpback whale entry in the year's Best 100 Photographs for the Museum's international exhibition. **Robyn Hughes**, who began diving in the 1980s, has travelled in almost sixty countries, and has become an accomplished wildlife and documentary photographer. Wade's and Robyn's recent *Looking for Whales* exhibition and book earned critical acclaim in Australian and international media. Their images hang in exclusive collections in the USA, Australia and Britain.

All other figures and other photos by the author with the exception of:

p. 17, fig. 1, courtesy Scott McVay

p. 42, fig. 2, courtesy George Crumb

p. 51, fig. 3, courtesy Sierra Club Books

pp. 58-59, orca mother with calf, © Slowmotiongli, Dreamstime.com

p. 93, fig. 6, courtesy Margaret Howe Lovatt

pp. 96-97, figs. 7 and 8, Sylvia van der Woude, "Effects of Signature Whistle Playbacks on Bottlenose Dolphins" (unpublished master's thesis, Berlin: Freie Universität Berlin, 2003). Used with permission of the author.

p. 98, fig. 9, Adam Pack, Louis Herman, Matthias Hofmann-Kuhnt, and Brian Branstetter, "The Object Behind the Echo: Dolphins Perceive Object Shape Globally Through Echolocation," *Behavioural Processes,* vol. 58, 2002, pp. 1–26. Used with permission of the authors.

pp. 112-113, friendly beluga whales, © Mandimiles | Dreamstime.com

p. 125, fig. 12, Courtesy of Rauno Lauhakangas and Jim Nollman.

pp. 149-154, figs. 17–23, Roger Payne and Scott McVay, "Songs of Humpback Whales," *Science,* vol. 173, no. 3997, August 13, 1971, pp. 585–597. Used with permission of the authors.

p. 157, figs. 26 and 27. Katherine Payne, Peter Tyack, and Roger Payne, "Progressive Changes in the Songs of Humpback Whales: A Detailed Analysis of Two Seasons in Hawaii," in Roger Payne, ed., *Communication and Behavior in Whales* (Boulder, CO: Westview Press, 1983), pp. 9–57. Used with permission.

pp. 164-165, figs. 28–29, Eduardo Mercado, Louis Herman, and Adam Pack, "Stereotypical sound patterns in humpback whale song," *Aquatic Mammals*, vol. 29, no. 1, 2003, pp. 37–52. Used with permission.

p. 166, fig. 30, Sean Green, Eduardo Mercado, Louis Herman, and Adam Pack, "Characterizing Patterns within Humpback Whale Songs," *Aquatic Mammals*, vol. 33, no. 2, 2007, pp. 202–213. Used with permission.

pp. 167-169, figs. 31–33, Nina Eriksen, Lee Miller, Jakob Tougaard, and David Helweg, "Cultural Change in the Songs of Humpback Whales from Tonga," *Behaviour*, no. 142, 2005, pp. 305–328. Used with permission.

p. 173, fig. 34, Andy Coghlan, "Whales Boast the Brain Cells That Make Us Human," *New Scientist*, November 27, 2006. Used with permission.

pp. 187, 190, figs. 36, 37, Hal Whitehead, *Sperm Whales: Social Evolution in the Ocean* (Chicago: University of Chicago Press, 2003). Used with permission.

pp. 204-205, a beached fin whale, © Imanolqs | Dreamstime.com

p. 214, fig. 38, 39, data compiled by Serge Masse.

pp. 218-219, figs. 41, 42. Erin Oleson, Sean Wiggins, and John Hildebrand, "Temporal Separation of Blue Whale Call Types on a Southern California Feeding Ground," *Animal Behaviour*, vol. 74, no. 4, 2007, pp. 881–894. Used with permission.

pp. 220-221, figs. 43, 44. Mark McDonald, Sarah Resnick, and John Hildebrand, "Biogeographic Characterization of Blue Whale Song Worldwide: Using Song to Identify Populations," *Journal of Cetacean Research Management*, vol. 8, no. 1, 2006, pp. 55–65. Used with permission.

p. 239, fig. 45, photo by Dan Sythe.

p. 256, fig. 50, photo by Olivia Wyatt, used by permission.

pp. 269-273, figs. 54-56, courtesy Michael Deal.

p. 279, fig. 58, courtesy Project CETI.

p. 280, fig. 59, courtesy Abigail Sanders.

p. 281, fig. 60, courtesy Alex South.

pp. 282-285, figs. 61-63, courtesy Annie Lewandowski and Kyle McDonald.

Musician and philosopher **David Rothenberg** wrote *Sudden Music, Why Birds Sing, Bug Music, Survival of the Beautiful, Nightingales in Berlin* and many other books, published in at least eleven languages. Born in 1962, he has more than thirty recordings out, including *One Dark Night I Left My Silent House* which came out on ECM, and most recently *In the Wake of Memories* and *Who Knows Why Whales Sing.* He has performed or recorded with Pauline Oliveros, Peter Gabriel, Ray Phiri, Suzanne Vega, Scanner, Elliott Sharp, Iva Bittová, and the Karnataka College of Percussion. His latest book is *Invisible Mountains.* Rothenberg is distinguished professor of philosophy and music at the New Jersey Institute of Technology.

www.davidrothenberg.net